THE STATUS O
NOTTINGHA

JASON REECE
Illustrations by CHRIS ORGILL

The time to see them is when night steals over the woods, and shadows flicker in the dying light, then the Fern Owls' jarring notes are heard.

Joseph Whitaker (1850-1932) describing the European Nightjar *Caprimulgus europaeus* in *Notes on the Birds of Nottinghamshire* 1907.

HOOPOE PRESS

JASON REECE lives in East Bridgford in south Nottinghamshire. He was educated at Nottingham, Nottingham Trent and Cambridge Universities and works as a barrister. He has been birding since 1981 and has seen 217 species in Nottinghamshire. He has written several articles about the birds of the county. This is his first book.

CHRIS ORGILL is a bird artist who lives and works at Chilwell in Nottinghamshire. His work has appeared in a number of field guides and other publications. He undertakes commissions and can be contacted on 0115 9229839.

This book is dedicated to Jeremy Robson who sowed the seeds for this work, to my wife Ila Reece who put up with it for so long and to the memory of Joseph Whitaker, W.J. Sterland and Austen Dobbs who pioneered the recording of birds in the county and of my grandmother Mary Ann Reece (1912-2008).

Published in the United Kingdom
by Hoopoe Press, 61 Kneeton Road, East Bridgford, Nottinghamshire NG13 8PG.

1st edition 2009

Copyright © 2009 text and tables by Jason Reece

ISBN: 978-0-9560592-0-8

Printed in the UK by Sherwood Print Services, Nottingham. Tel 0115 9208612

All rights reserved. No part of this publication may be reproduced, stored in a retrieval system or transmitted in any form or by any means, electronic, mechanical, photocopying, recording or otherwise, without the prior written permission of the copyright owner.

British Library Cataloguing in Publication Data. A cataloguing record for this book is available from the British Library.

CONTENTS

Introduction	*page 4*
The County of Nottinghamshire	*pages 5-6*
Birding in Nottinghamshire	*pages 6-12*
Key sites	*pages 13-15*
Significant events 1974-2007	*pages 16-17*
Species list	*pages 18-135*
Other records	*pages 135-138*
Appendix 1 Records of County Rarities	*pages 139-141*
Appendix 2 First and Last dates for migrants	*pages 142-146*
Bibliography	*pages 147-148*
Full county checklist	*pages 149-156*
Index	*page 156*

INTRODUCTION

The first major assessment of the birds of Nottinghamshire was written by the great Victorian ornithologist Joseph Whitaker in 1907. The last one hundred years have witnessed enormous changes in life in Nottinghamshire and 2009 would therefore seem to be an appropriate date for a fresh assessment of the birds of the county.

This book is intended to be a brief introduction to the birds of Nottinghamshire and summarises the status of each species which has been recorded in the county up to 2007. It is also intended to supplement the last comprehensive county avifauna written by Austen Dobbs in 1975 and to analyse the patterns of occurrence of each species within the new county boundaries established in 1974.

With regard to controversial records I have not sought to act as a one-man record committee but have adopted those records accepted elsewhere by the British Birds Rarities Committee and by Nottinghamshire Birdwatchers without further comment.

I would like to thank Chris Orgill for providing the line drawings which accompany the text. I would also like to thank those members of Nottinghamshire Birdwatchers - particularly Bernie Ellis, Andy Hall, Phil Palmer and Michael Warren - who encouraged this project and helped to check the draft text.

Any errors or omissions are - of course - mine but I would encourage any observers with details of records not included here to submit details to Nottinghamshire Birdwatchers so that important records are not lost to posterity.

Jason Reece

East Bridgford, Nottinghamshire

April 2009

THE COUNTY OF NOTTINGHAMSHIRE

Nottinghamshire is very much a north midland county marking the boundary between the English Midlands and the true North as may be seen from the counties which surround it - Leicestershire to the south and south-east, Derbyshire to the west, Lincolnshire to the east and north-east and South Yorkshire to the north-west.

BOUNDARY CHANGES
In one significant way Nottinghamshire is an easier county to analyse than many other English and Welsh counties since the boundaries of the county have not been altered to any great extent by the various local government boundary reviews which have taken place since the nineteenth century. In that sense at least the Nottinghamshire of today would be familiar to Joseph Whitaker and it is therefore possible to make direct comparisons between his 1907 work and modern bird records. In total only four minor amendments to the county boundary have been made since 1895. In the west of the county a part of Kirkby-in-Ashfield and the bulk of Pinxton were moved to Derbyshire in 1895. Subsequently part of Willoughby-on-the-Wolds in the south of the county was moved to Leicestershire in 1965. Finally in 1974 Finningley and part of Harworth was moved into Yorkshire in the major reorganisation of 1974. Since then the boundary has not been altered.

TOPOGRAPHY
Nottinghamshire has few significant landscape features and is primarily a lowland county rising to between 500 and 651 feet above sea level in the west in the Huthwaite - Moorgreen area. From north to south the county is 52 miles long and runs 27 miles east to west at its widest extent. In total the county occupies just short of 850 square miles.

The dominant landscape feature is the River Trent which enters the south-west of the county to the west of Nottingham and runs diagonally to the north-east leaving the county beyond Newark. The majority of the county drains into the Trent and the Trent Valley is a significant route for migrating birds.

Most of the standing water in the county has been created by human activity and occupies three main areas - the large man-made lakes in the Dukeries (the largest being Welbeck Great Lake at 89 acres and Clumber Lake at 83 acres), the various flooded gravel pits along the Trent Valley and at Lound / Hallcroft and the single large reservoir at King's Mill (70 acres). The county therefore lacks enormous modern reservoirs like Draycote Water in Warwickshire and Chew Valley Lake in Somerset. There has also been a lack of decent wader habitat in the county in recent years as many gravel pits have been flooded and margins have become overgrown. One particularly significant wader habitat - the 1000 acres of marshland at Nottingham Sewage Farm east of the city - was lost through redevelopment in 1960-61.

The western half of the county is underlain by a bed of bunter sandstone running north to south for 40 miles and up to 8 miles wide in places. This has created a series of significant habitats - particularly between Mansfield and Worksop - comprising precious fragments of lowland heath (for example at Budby) and large areas of woodland dominated by conifer production (as at Sherwood Pines). Beyond Ollerton the woodland becomes more mixed and there are significant tracts of old deciduous woodland at Edwinstowe and in the great parks of the Dukeries.

In the far north and most of the south and eastern side of Nottinghamshire the county is essentially agricultural. The south and eastern farmland is now primarily given over to well managed arable production. In contrast the northern Carrlands in the Idle Valley retain a less intensive character with pockets of birch, willow and hawthorn scrub and large fields split by deep dykes and marked out by untidy margins. As a result many of the more sensitive farmland birds such as Common Quail and Corn Bunting still breed in the north, having virtually vanished elsewhere.

Finally the urban and industrial imprint of the county has to be considered. The total population of the county was estimated at 1,055,400 people in 2006. The major centre of population is in the south west of the county concentrated on the city of Nottingham and supporting commuter communities in the boroughs of Broxtowe, Gedling and Rushcliffe. To the west of Nottingham a series of small towns run along the border with Derbyshire north to Mansfield, following the line of the coal measures from which their income formerly derived. Beyond Mansfield the town of Worksop is the only large settlement in the north-west, whilst the market towns of East Retford and Newark are the key towns in east Nottinghamshire.

BIRDING IN NOTTINGHAMSHIRE

SOURCES
There are five main published sources of bird records for Nottinghamshire. Two of these are works covering the whole county by Joseph Whitaker (*Notes on the Birds of Nottinghamshire* Nottingham 1907) and by Austen Dobbs (*The Birds of Nottinghamshire Past and Present* Newton Abbot 1975). In addition John Knifton's *Rare Birds in Nottinghamshire 1759-1992*

Nottingham 1992 and Keith Naylor's *The Rare and Scarce Birds of Nottinghamshire* Nottingham 1996 are key sources for records of county rarities. The fifth major source of records is the annual reports of the county bird club Nottinghamshire Birdwatchers (formerly the Trent Valley Bird Watchers) produced from 1943 onwards. As will be seen from the species accounts which follow, the club - which was formed in 1935 - is an active one carrying out regular surveys of birds within the county as well as running a lively programme of indoor and outdoor meetings. Other significant sources of records are listed in the bibliography.

THE COUNTY LIST

The current Nottinghamshire list stands at 325 species.

Using the same modern list of British species recognised by the British Ornithologist's Union Whitaker recorded a county list of 250 species in 1907, setting aside some old records which are no longer accepted.

Austen Dobbs listed records for 287 species for the period to 1975.

In total 300 species were recorded between 1974 and 2007.

The most significant developments over the last two centuries are set out below:

Firsts for Britain recorded in Nottinghamshire.

SPECIES	LOCALITY	DATE
Redhead	Bleasby & Gibsmere	08-27.03.1996.
Lesser Yellowlegs	Misson	Winter 1854-55.
Egyptian Nightjar	Thieves Wood, Mansfield	23.06.1883.
Dusky Thrush	near Gunthorpe	13.10.1905

Additions to the Nottinghamshire List 1800-2007.

YEARS	ADDITIONS	CULMULATIVE TOTAL
1800-1949	NA	267
1950-59	3	270
1960-69	10	280
1970-79	14	294
1980-89	7	301
1990-99	17	318
2000-07	7	325

Species unrecorded in Nottinghamshire after 1973.

YEAR OF LAST RECORD	NUMBER OF SPECIES	DETAILS OF LAST RECORD
1800-1899	7	Woodchat Shrike (1859) Little Bustard (1866) Two-barred Crossbill (1875) Egyptian Nightjar (1883) Nutcracker (1883) Pallas's Sandgrouse (1889) Pine Grosbeak (1890)
1900-09	3	Dusky Thrush (1905) Great Bustard (1906) Glossy Ibis (1909)
1910-19	1	Black Grouse (1917)
1920-29	2	Little Bittern (1921) Baillon's Crake (1922)
1930-39	0	
1940-49	1	Icterine Warbler (1945)
1950-59	4	Short-toed Lark (1950) Little Bunting (1950) Cirl Bunting (1953) Rose-coloured Starling (1959)
1960-69	4	Broad-billed Sandpiper (1961) Wilson's Phalarope (1961) Solitary Sandpiper (1962) European Roller (1966)
1970-73	3	European Bee-eater (1970) Common Nighthawk (1971) Eurasian Scops Owl (1973)

Lost breeding species in Nottinghamshire 1800-2007.

YEARS	LOST BREEDING SPECIES	LAST BREEDING RECORD
1800-1899	Red Kite	early 19th century
	Merlin	19th century
	Twite	19th century
	Wryneck	19th century
	Stone-Curlew	1891
1900-1949	Black Grouse	c1910
	Cirl Bunting	1940s
1950-1999	Corncrake	1968
	Northern Wheatear	1970
	Red-backed Shrike	1977
	Wood Warbler	1996
2000-2007	Whinchat	2007
	12 lost breeding species	

New breeding species in Nottinghamshire 1800-2007.

YEARS	NEW BREEDING SPECIES	FIRST BREEDING RECORD
1800-1899	Canada Goose	19th century
	Tufted Duck	1830s-1840s
	Red-legged Partridge	c1850s
	Turtle Dove	1868
	Shoveler	1874
	Common Redshank	c1884
1900-09		
1910-19	Little Owl	1913
1920-29	Common Shelduck	1921
	Black-headed Gull	1928
1930-39		
1940-49	Eurasian Curlew	c1944
	Garganey	1945
	Common Pochard	1945
	Common Tern	1945
1950-59	Grey Wagtail	1955
	Little Ringed Plover	1956
	Collared Dove	1959
1960-69	Honey-Buzzard	1960s
	Common Crossbill	1967
	Gadwall	1968
1970-79	Greylag Goose	1971
	Water Rail	1971
	Ringed Plover	1974
	Black-necked Grebe	1979
	Hobby	1979
	Oystercatcher	1979
1980-89	Ruddy Duck	1980
	Northern Goshawk	1980s
1990-99	Great Cormorant	1990
	Common Sandpiper	1990
	Siskin	1990
	Barnacle Goose	1990s
	Eurasian Wigeon	1997
	Egyptian Goose	1998
2000-07	Mandarin Duck	2000
	Red-crested Pochard	2001
	Peregrine Falcon	2004
	Cetti's Warbler	2007
	37 new breeding species	

LITTLE OWL
A species which was first recorded in Nottinghamshire in 1896 and first bred in the county in 1913.

Occasional breeding species in Nottinghamshire 1800-2007

SPECIES	BREEDING DETAILS
Ring Ouzel	1856
Spotted Crake	c1871
Black-winged Stilt	1945
Common Gull	1967-69
Marsh Warbler	1969
Pied Flycatcher	3 occasions - last 1998
Short-eared Owl	4 occasions - last 1999
Black Redstart	Several occasions - last 2003
Lesser Black-backed Gull	3 occasions - last 2007

OTHER BREEDING RECORDS

Black Tern (1978) and Little Gull (1987) have both attempted to breed in Nottinghamshire.

Pallas's Sandgrouse (1888), Montagu's Harrier (1955-56), Pintail (1969) and Fieldfare (1984) have all been suspected of breeding at least once.

In addition a feral pair of White-fronted Geese bred in Nottinghamshire in 2001.

The number of species recorded annually (including feral geese and Ruddy Shelduck) and the number of observers submitting records in the 34 years between 1974 and 2007 is as follows:

YEAR	SPECIES	OBSERVERS	YEAR	SPECIES	OBSERVERS
1974	186	235	1991	204	192
1975	186	241	1992	200	191
1976	204	222	1993	205	203
1977	205	251	1994	222	176
1978	200	233	1995	207	164
1979	201	240	1996	223	213
1980	194	219	1997	215	233
1981	194	229	1998	214	200
1982	191	235	1999	207	216
1983	205	216	2000	213	206
1984	198	250	2001	216	219
1985	196	250	2002	215	253
1986	206	237	2003	216	249
1987	194	191	2004	212	175
1988	209	179	2005	215	178
1989	210	212	2006	218	161
1990	215	212	2007	213	-

PREDICTIONS

It is difficult to speculate which species will be added to the county list. There have been some spectacular rarities in the English Midlands in recent years which would not have been predicted by anyone such as the Belted Kingfisher *Megaceryle alcyon* in Staffordshire in April 2005. However a comparison with the lists for other Midlands counties makes clear that there are plenty of potential vagrants which have been recorded in neighbouring counties:

COUNTY	LIST*	YEAR	SPECIES UNRECORDED IN NOTTINGHAMSHIRE
Derbyshire	311	2006	16 species
Leicestershire & Rutland	307	2005	15 species
Northamptonshire	318	2003	19 species
Nottinghamshire	325	2007	Not applicable
Staffordshire	311	2006	18 species
Warwickshire	307	2006	16 species
Worcestershire	297	2006	13 species
West Midlands	270	2006	3 species
Total	**378**	**NA**	**53 species**

* *Escaped waterfowl and Golden Pheasant are excluded from the totals.*

The most likely additions would seem to fall into the following categories -

British breeding birds (1 species)
Dartford Warbler *Sylvia undata*.

Scarce British coastal migrants (4 species)
Tawny Pipit *Anthus campestris*, Aquatic Warbler *Acrocephalus paludicola*, Melodious Warbler *Hippolais polyglotta*, Common Rosefinch *Carpodacus erythrinus*.

Regular Vagrants to Britain (12 species)
Pied-billed Grebe *Podilymbus podiceps*, Collared Pratincole *Glareola pratincola*, Black-winged Pratincole *Glareola nordmanni*, White-rumped Sandpiper *Calidris fuscicollis*, Least Sandpiper *Calidris minutilla*, Marsh Sandpiper *Tringa stagnatilis*, Laughing Gull *Larus atricilla*, Franklin's Gull *Larus pipixcan*, Bonaparte's Gull *Chroicephalus philadelphia*, Citrine Wagtail *Motacilla citreola*, Black-throated Thrush *Turdus ruficollis*, Lesser Grey Shrike *Lanius minor*.

Turning to breeding birds it will be apparent from the species accounts which follow that several species would seem to be on the verge of either colonising or returning to Nottinghamshire as breeding birds, notably Little Egret, Red Kite, Marsh Harrier, Osprey, Lesser Black-backed Gull, Cetti's Warbler, Common Raven and perhaps Pied Flycatcher. They may be followed as regular breeding birds by Avocet, Mediterranean Gull, Herring Gull and Firecrest.

On the debit side several species are in serious decline or have a restricted distribution within the county and may be lost in the years ahead unless trends are reversed, particularly Eurasian Teal, Honey-Buzzard, Common Snipe, Eurasian Curlew, Common Nightingale, Black Redstart, Common Redstart, Lesser Redpoll and perhaps Corn Bunting. The creation of stable, managed wetland would assist threatened species of wading birds in particular.

LISTING
There are a number of active listers within the county and a number of observers have seen in excess of 200 species within the current county boundaries. At the time of writing the leading published county lister is John Hopper who has seen over 270 species in Nottinghamshire.
Phil Palmer uniquely recorded in excess of 200 species in the county in one calendar year (1996) -*see NBAR 1995 pp.117-120 and 1996 pp.105-107.*

KEY SITES (with grid references)

No right of access is implied in the site accounts listed below.

TRENT VALLEY (south to north).

Attenborough Nature Reserve (SK522344) - gravel pit complex. Reserve centre at Barton Lane off the A6005 through Chilwell Retail Park.
Colwick Country Park (SK610395) - landscaped gravel pits and parkland. Off the A612 east of Nottingham.
Holme Pierrepont (SK628394) - gravel pits and scrub. Viewable from the A52 and the access road to Holme Pierrepont Hall.
Netherfield Lagoons (SK638402) - slurry lagoon and other pits. From the A612 east of Nottingham through Victoria Retail Park.
Burton Meadows (SK660434) and **Gunthorpe** (SK678727) - gravel pits and farmland. Off the A6097 south of Gunthorpe and at Meadow Lane, Burton Joyce.
Hoveringham (SK715474 & SK696483) - gravel pits. Off the A612 at Hoveringham / Thurgarton.
Bleasby (SK715495) - a private gravel pit complex. From Bleasby village.
Langford Lowfields (SK815615) - gravel pits being developed by RSPB. Off the A1133 through Collingham village and down Westfield Lane.
Collingham Pits (SK829614) - new gravel pits. Off the A1133 through Collingham village and down Carlton Ferry Lane.
Girton (SK818673) - gravel pits. Off the A1133 along Trent Lane and other minor roads north of Girton.
Cottam (SK815798) - Lagoons and scrub. Footpath from Laneham.

RIVER IDLE CATCHMENT

Gringley Carr (SK7293) - **Idle Stop** (SK721965) - **Misson Carr** (SK695951) - open farmland and carrland. Minor roads from Gringley-on-the-Hill and nearby villages.
Lound GPs (SK693857) - gravel pits. Limited access from Lound village at Chainbridge Lane or on foot at Neatholme Lane. **Hallcroft GPs** (SK692832) are nearby. The sites will form **Idle Valley NR** in 2008 accessed from Hallcroft / Hayton.

NORTH WEST WOODS & HEATHS

Clumber Park (SK626746) - woodland, heathland and lake belonging to the National Trust. From the A614 between Worksop and Ollerton.
Budby Heath / Common (SK615690) - lowland heath. Off the A616 between Ollerton and Cuckney.
Sherwood Forest Country Park (SK627677) / **Birklands** (SK622677) - ancient woodland. Off the B6034 Edwinstowe to Budby road.
Clipstone Forest (SK616626) - conifer plantations. Car park at Sherwood Pines Forest Park off the B6030 Clipstone Road.
Rufford CP (SK647654) - woodland, parkland and small lake. From the A614 south of Ollerton.
Welbeck Raptor Watchpoint (SK581721) - views across Welbeck estate. From the minor road between Carburton and Cuckney.

OTHER SITES

Cotham Flash (SK795494) - small shallow pools. Off the A46 at Farndon turning to Hawton and then south towards Cotham village.
Kilvington New Lake (SK795433) - flooded gypsum pit with muddy margin. North from the A52 at Elton-on-the-Hill.
King's Mill Reservoir (SK516594) - large reservoir. On the A38 west of Mansfield.
Annesley Pit Top (SK522538) - brownfield pit top site with pools. From the railway crossing in Annesley village.
Blidworth Forest (SK591545 & SK591549) - pine woods. From minor roads -Blidworth Lane and Longdale Lane - off the A614 north-west of Calverton.
Bennerley Marsh (SK474438) - river meadows on the Derbyshire border. View from minor road between Awsworth and Ilkeston.
Erewash Meadows NR (SK452484) - Also known as Langley Mill and Brinsley Flashes - floodplain, pools and farmland. From the A610.
Wollaton Park (SK532393) - large parkland and woodland. Off the A609 Wollaton Road west of Nottingham city centre.
Nottingham General Cemetery (SK566403) - Victorian cemetery. On foot from Nottingham city centre at Canning Circus.
Bunny Old Wood (SK585284) - ancient elm woodland. Off the A60.
Eakring Flash (SK676628) - subsidence pools. Off the A614 through Eakring village. The ancient woodland at **Duke's Wood** (SK682603) is nearby.

KEY SITES IN NOTTINGHAMSHIRE

SIGNIFICANT EVENTS 1974-2007

1974	First Dark-breasted Barn Owl. Ringed Plover bred successfully for the first time.
1975	First Buff-breasted Sandpiper. Great Crested Grebe Survey. Rook Survey.
1976	First Great Reed Warbler. Common Nightingale Survey.
1977	
1978	First Black Kite & Sociable Lapwing. Breeding attempt by Black Tern. Mute Swan Survey.
1979	First Mediterranean Gull & Bluethroat. Black-necked Grebe bred for the first time. Hobby proved to breed for the first time. Oystercatcher proved to breed for the first time. Record numbers of Red-necked Grebe, Black-throated Diver, Red-throated Diver and Red-breasted Merganser (February-March). Common Kestrel Survey (to 1981).
1980	Ruddy Duck bred for the first time. Common Nightingale Survey.
1981	First Killdeer, Yellow-legged Gull & Cetti's Warbler. European Nightjar Survey.
1982	Common Snipe & Common Redshank Survey.
1983	First Ring-necked Duck and Scandinavian Rock Pipit. Influx of Shag (February). Mute Swan Survey.
1984	First American Golden Plover. Influx of Little Gull (May). Little Ringed Plover Survey.
1985	
1986	First 'Siberian' Chiffchaff.
1987	First Arctic Redpoll. Breeding attempt by Little Gull.
1988	Influx of Kittiwake (March).
1989	First Green-winged Teal.
1990	First Lesser Scaup, Ring-billed Gull, Savi's Warbler & Parrot Crossbill. Great Cormorant bred for the first time. Siskin proved to breed for the first time. Common Sandpiper proved to breed for the first time.

	Influx of Black Tern (May). Mute Swan Survey.
1991	
1992	
1993	First European Serin. Influx of Common Eider (October-November).
1994	First Bufflehead, Spotted Sandpiper, Red-rumped Swallow, Yellow-browed Warbler & Penduline Tit.
1995	Influx of Arctic Redpoll & Mealy Redpoll (December to April 1996). Barnacle Goose bred for the first time.
1996	First Redhead, Long-billed Dowitcher & Cedar Waxwing. Rook Survey. Influx of Bean Goose (January - February). Influx of Waxwing (January - April).
1997	First Caspian Gull.
1998	First Baird's Sandpiper & Alpine Swift. Egyptian Goose bred for the first time. Influx of Arctic Tern (May). Turtle Dove Survey.
1999	First Cattle Egret. Exceptional passage of Pink-footed Goose (January). Influx of Little Gull (April). Little Ringed Plover & Tree Pipit Survey.
2000	First Blue-winged Teal. Mandarin Duck proved to breed for the first time. Influx of Honey-Buzzard (September-October). Reed Bunting Survey.
2001	First Little Swift.
2002	First Whiskered Tern, Blyth's Pipit & Pallas's Leaf Warbler.
2003	
2004	First Sora. Peregrine proved to breed for the first time. European Nightjar Survey. Spotted Flycatcher Survey.
2005	Exceptional passage of Pink-footed Goose (January). Influx of Waxwing (January-May).
2006	First Kumlein's Gull.
2007	First Dusky Warbler. Cetti's Warbler bred for the first time.

SPECIES LIST

Records are analysed in detail for the period 1974-2007 at which time the modern boundaries of the county were established. The species accounts therefore supplement the data listed by Austen Dobbs in *The Birds of Nottinghamshire Past and Present* for the period up to 1973. In many instances a brief summary of the status of the species before 1974 has been included in the account for each bird.

Tabular records for each species are all based upon arrival dates for individual birds.

With regard to extreme rarities all records are given for those species which have been recorded in the county on ten occasions or less and for several other scarce species. Only those records officially accepted and published by the British Birds Rarities Committee (BBRC) or - in the case of local rarities - by Nottinghamshire Birdwatchers are set out in the individual species accounts. As a result the records for the rarest species mostly accord with those listed by K.A. Naylor in 1996 with any significant discrepancies listed in the text.

For lesser rarities the records broadly accord with those published in the Nottinghamshire Birdwatchers Annual Reports and listed by K.A. Naylor with one or two minor corrections for obvious errors or rejected records. Where any discrepancies exist between records listed in Naylor and the Nottinghamshire Birdwatchers Annual Report I have preferred the records in the Annual Report in most instances.

With regard to the abundance of each species in Nottinghamshire a slightly modified version of the definitions used by the Nottinghamshire Birdwatchers has been adopted in this work so far as possible:

Very rare	10 or fewer county records.
Rare	11 to 50 county records.
Scarce	51 to 200 county records.
Uncommon	annual or virtually annual with up to 20 records per year.
Fairly common	likely to be seen during most visits to suitable habitats (but usually in relatively small numbers).
Common	likely to be seen in reasonable numbers on all visits to suitable habitat.

KEY ABBREVIATIONS

BBRC	British Birds Rarities Committee.
CP	Country Park.
GP	Gravel Pit.
NBAR	Nottinghamshire Birdwatchers Annual Report
NR	Nature Reserve.
PS	Power Station.
RW	Raptor Watchpoint.
SF	Sewage Farm.
♂	Male.
♀	Female.

☐ MUTE SWAN *Cygnus olor*

A reasonably common and conspicuous resident species. A survey in 1955 revealed c70 breeding pairs (276 adults) in Nottinghamshire. Numbers declined in the 1960s and 1970s as many birds died from ingesting lead fishing weights. Further surveys revealed 58 pairs in 1978 and 48 pairs and 193 birds in 1983. Since legislation was introduced to restrict the use of lead weights in 1987 numbers have recovered. A further survey 3 years later found 364 birds in the county.

Mute Swan - maximum site counts in Nottinghamshire 1975-2007

Years	1975-79	1980-84	1985-89	1990-94	1995-99	2000-04	2005-07
Site maxima	107	117	67	130-140	203	217	241

Lound has become the key site for this species since the 1990s. 30 pairs bred there in 2001 and a county record of 241 birds were in the Lound/ Clayworth Common area on 17.12.2006. There are now also good numbers in the lower Trent Valley and around 700 now winter in the county.
See NBAR 1990 p.64.

☐ BEWICK'S SWAN *Cygnus columbianus*

Formerly a scarce species in Nottinghamshire, the Bewick's Swan is now a regular winter visitor in small numbers, principally to the north Nottinghamshire Carrlands/Idle Valley. Numbers peaked in the period 1976-1980 and have declined since then to the point that Whooper Swan has become the commoner visitor to the county as shown by the table below (when compared to that for Whooper Swan):

Bewick's Swan - maximum site counts in Nottinghamshire 1975-2007

Years	1975-79	1980-84	1985-89	1990-94	1995-99	2000-04	2005-07
Annual site maxima	14-111	27-120	35-73	34-80	42-99	16-39	35-51

The largest count for 1974-2007 was a herd of 120 on floods at Misson 26.02.1980.
Extreme dates 04.10.1998 (Lound) - 16.05.1987 (an injured bird at Lound).

☐ **WHOOPER SWAN** *Cygnus cygnus*
Prior to 1974 this species was a scarce and somewhat irregular winter visitor to Nottinghamshire. This remains the picture today in most of the county but small numbers (20-50 birds) now winter in the north Nottinghamshire Carrlands/Idle Valley (for example at Misson and Gringley) each year. Numbers have built up steadily as shown below:

Whooper Swan - maximum site counts in Nottinghamshire 1975 -2007

Years	1975-79	1980-84	1985-89	1990-94	1995-99	2000-04	2005-07
Annual site maxima	3-20	9-26	8-32	11-35	9-36	23-73	68-73

The largest counts for 1974-2007 were of 73 over Papplewick on 13.03.2002 (with 4 other flocks with a combined total of 71-95 birds moving through the county on the same date) and 73 in the Idle Valley on 12.03.2006. Birds usually arrive in late November and are mostly gone by late March.
Extreme dates 20.09.1995 (Bennerley Marsh) - 16.06.2007 (Kilvington New Lake). In addition one bird summered at Welbeck 02.07-08.10.1949.

☐ **BEAN GOOSE** *Anser fabalis*
The Bean Goose is a very scarce winter visitor to Nottinghamshire and the rest of the English Midlands. Two races occur in the county - the **Taiga Bean Goose** *A. f. fabalis* and the **Tundra Bean Goose** *A. f. rossicus* - although many records have not been separated at a sub-specific level.
There is only one record involving wild birds before 1981 - 7 birds at Annesley in December 1890 (2 of which were shot).
There were records of presumed or suspected wild birds in 10 years between 1974 and 2007:
1981   37 Barton Fields 19.12.1981.
1982   7 Besthorpe 23.01.1982, 30 Attenborough NR 04.03.1982.
1990   10 Colwick CP 29.01.1990.
1991   4 Holme Pierrepont 13.01.1991.
1995   1 Lound late December 1995 (*Tundra*).
1996   An influx of up to 62 birds in January - February with 3 Holme Pierrepont 21-27.01.1996, 9 Idle Valley 28.01-02.02.1996, 16 Lound 28.01-03.02.1996, 14 Grassthorpe 03.02.1996. 10 Girton 10.02.1996, 5 between Gringley and Misterton 10.02.1996 and 5 Lound 10-11.02.1996.
Later the same year - 7 Idle Valley 27.12.1996, 5 Besthorpe 30.12.1996.

1997 2 Lound 02.01-03.02.1997 (*Tundra*), 4 Misterton 02-08.02.1997 (*Tundra*),
15 Colwick 30.12.1997 (*Taiga*).
2001 2 Burton Meadows 01.01.2001 (*Tundra*).
2003 10 Gringley Carr 03-04.01.2003 (*Tundra*), 2 Lound 10.02.2003.
2007 1 Bentinck Pit Top 15.12.2007.

As with several other species of goose a small number of escaped birds of both races have also been recorded with between 1 and 5 birds in most years since 1979.

◻ PINK-FOOTED GOOSE *Anser brachyrhynchus*

There has been an enormous increase in passage numbers of this goose through Nottinghamshire which reflects the massive rise in wintering numbers in Britain. The county is on the flight line for birds moving between key wintering sites on the Lancashire Mosses and in North Norfolk.

No wild birds were reported at all in 1974 but since the late 1970s large movements of Pink-footed Geese have taken place in spring and autumn.

Pink-footed Goose - maximum daily movements over Nottinghamshire 1975-2007

Years	1975-79	1980-84	1985-89	1990-94	1995-99	2000-04	2005-07
Maximum movement	350	700	1320	1000	7400	4490	7910

The largest passage movements were in 1999 (with over 30,000 birds reported through the county including 17,000 birds passing through in January alone) and in January 2005 (when 20,000 birds went through the county). 7910 passed through Nottinghamshire on 23.01.2005 and 12 skeins totalling 4000 birds went through Lound on 22.01.1999 - the best site count for 1974-2007.

◻ WHITE-FRONTED GOOSE *Anser albifrons albifrons*

An uncommon winter visitor and passage migrant through Nottinghamshire. There were records of presumed wild birds in 26 of the 34 years between 1974 and 2007 but many records only involved skeins flying through the county.

White-fronted Goose - birds recorded in Nottinghamshire 1975-2007

Years	No. of years recorded	Birds January - April	Birds October - December	Maximum count
1975-79	3	83	11	74
1980-84	2	31	7	10
1985-89	5	71-72	46	38
1990-94	3	79	43	33
1995-99	5	408-409	165	150
2000-04	5	68	373	200
2005-07	2	0	85	68

December and January are the key months for records and there were 2 significant movements in the period - 150 west at Besthorpe 15.01.1997 and 200 through Attenborough and Toton 16.12.2000 (with a further 100 over Toton later the same day).
Extreme dates 02.10.2007 (Lambley) - 02.05.2000 (Hoveringham).
Small numbers of feral birds have also been recorded with increasing regularity since at least 1977 with 1 pair breeding at Wheatgrass Farm, Chilwell in 2001.
In addition to the birds listed in the table above **Greenland White-fronted Geese A. a. flavirostris** have been recorded in the county at least 4 times - 1 at Lound 14.04.1973, 9 at Misson 23.12.1979, 1 at King's Mill Reservoir 04-08.01.1986 and then at Lound 18.01-06.03.1986 and 1 at Hoveringham 19.01.2003. There are no regular wintering flocks of this subspecies anywhere in England.

□ GREYLAG GOOSE *Anser anser*
Historically this bird was a scarce and irregular wild visitor to Nottinghamshire. Today the picture is totally dominated by feral birds. The species first bred in 1971 and there were only 6 breeding pairs as late as 1985. Since then numbers have soared with the first three figure flock in 1990, the first 500+ count in 1996 and the first 1000+ count in 2001.
Lound is a key site for this species with 125 breeding pairs in 2001 and the maximum count for Nottinghamshire is of 2000 there on 16.01.2006. The highest count for the rest of the county is of 1254 birds in the Trent Valley at Shelford/Burton Joyce on 17.09.2006.
The Greylag Goose is still increasing in Nottinghamshire and the county population exceeds 3000 adult birds today.

□ CANADA GOOSE *Branta canadensis*
This introduced bird has bred in Nottinghamshire since the nineteenth century and it was locally fairly common in the county by 1974 with flocks of up to 450-500 birds recorded.
Since 1974 numbers have built up considerably and the species is now common on most waters in the county. This is shown by monthly county wildfowl counts of 1010 (September 1976), c.2000 (September 1979), 3385 (September 1986) and 5360 (September 2000). Netherfield alone held 1974 post-breeding birds on 15.09.1996 and 2000 on 04.09.2000 and numbers may still be increasing in Nottinghamshire.

□ BARNACLE GOOSE *Branta leucopsis*
The Barnacle Goose is primarily a rare feral bird in Nottinghamshire and 1 to 6 escaped birds were recorded annually between 1974 and 1993. Numbers of feral birds have increased significantly since then and there is now a small population in the Osberton area where breeding was first recorded in 1995. 70 birds were recorded at the Osberton Estate on

30.09.2007 and birds from this site presumably account for records of 50 birds at Clumber 05.01.2002 and 42 birds at the same site 12.12.2005. Suspected wild birds are extremely rare. There were 3 such records in the 19th century - 52 over Ramsdale in September 1869, 1 shot at Eastwood 13.12.1890 and an undated 19th century record of 1 shot at Mansfield. Subsequently there were 8 records between 1974 and 2007 which may have involved wild birds (although some of these records possibly also involved feral birds, particularly the birds at Rufford in 1981 and at Girton in 1993):

1981   4 Rufford 03.02.1981.
1986   10 north-east at Clipstone Forest 27.01.1986.
1988   7 Daneshill Reservoir 03-06.12.1988.
1993   4 Girton 28.11.1993.
1994   7 Hoveringham 18.02-07.03.1994.
1996   6 Girton 23-26.03.1996.
2003   30 Collingham Pits 28.12.2003.
2006   31 north-west over Clumber 13.01.2006.

☐ **BRENT GOOSE** *Branta bernicla bernicla*
Historically this species was a very rare straggler to Nottinghamshire with just 6 records between 1850 and 1973. Since 1974 it has been recorded in the county far more frequently as the number of birds wintering around the British coast has increased. However records remain less than annual with 11 blank years to 2007.

Brent Goose - birds recorded in Nottinghamshire 1975-2007

Years	1975-79	1980-84	1985-89	1990-94	1995-99	2000-04	2005-07	Total
Birds	47	8	31	18	56	11	6	177

Months	J	F	M	A	M	J	J	A	S	O	N	D
Birds	28	7	32	7	0	0	0	0	0	23	52	28

Only 2 years in the period recorded more than 14 birds (1979 with 46 birds and 1996 with 30 birds) and only 3 double figure flocks were recorded - 12 at Bulcote 18.03.1979, 12 at Colwick CP 21.10.1987 and 33 through Oxton Bogs 11.11.1979.
Most birds are of the **Dark-bellied race** *B. b. bernicla* but 1 **Pale-bellied Brent Goose** *B. b. hrota* was at Colwick 20.10.1988 and 2 were at Hoveringham 29.03.1999 (and are included in the tables above).
Extreme dates 11.10.1888 (River Erewash and River Trent) & 11.10.1992 (Stoke Bardolph) - early June 1992 (Lound).

☐ **EGYPTIAN GOOSE** *Alopochen aegyptiaca*
There are a few old records of escaped Egyptian Geese in Nottinghamshire back to 1868 when one was shot at Eastwood. Since 1973 a feral population has gradually become established in the county as elsewhere in

England. From 1973 to 1979 a maximum of 5-6 released birds were seen annually, primarily in the Trent Valley where a pair summered at Hoveringham in 1974 and a pair was present at Attenborough NR in 1975 and attempted to breed at the latter site in 1976. This introduction failed with only 4 birds recorded between 1980 and 1986. Since then numbers have built up again, particularly after further releases in the late 1990s.

Egyptian Goose - definite breeding records for Nottinghamshire 1998-2005

Years	Breeding details
1998	Rufford (1 escaped pair).
1999	Centre Parcs (1 pair).
2000	Attenborough NR (1 pair). Centre Parcs (1 pair).
2001	Attenborough NR (1 pair). Centre Parcs (2 broods).
2002	5 pairs bred at 4 sites.
2003	6 pairs bred at 3 sites.
2004	Breeding at 5 sites.
2005	7 pairs bred at 3 sites.

There are now at least 2 breeding populations in Nottinghamshire - one around Clumber, Rufford and Centre Parcs (where 5 pairs bred in 2007) and the other in the lower Trent Valley between Attenborough NR and Hoveringham - and further consolidation may be expected.
20 were counted at Attenborough NR 06.08.2003 and the county population is currently in the region of 50 birds.

□ **RUDDY SHELDUCK** *Tadorna ferruginea*
There is one old Nottinghamshire record of 2 Ruddy Shelduck at Newstead Abbey in 1869 which might relate to wild birds. However the British Ornithologists Union accepts no wild records for Britain before 1892.
Further records followed in the 20th century with 2 at Bulcote 12.09.1943, 1 near Netherfield 17.09.1944 and 1 at Hoveringham 28.05-27.06.1973.
More recently, what are presumed to be escaped birds have been recorded on several occasions with 8 together at Hoveringham 18.09.1983. There is a clear peak in records between June and September.

Ruddy Shelduck - birds recorded in Nottinghamshire 1975-2007

Years	1975-79	1980-84	1985-89	1990-94	1995-99	2000-04	2005-07	Total
Birds	1	10-12	2	1	3	9-10	0	26-29

Months	J	F	M	A	M	J	J	A	S	O	N	D
Birds	1	0	0	1	3	8-10	1-2	3	8	1	0	0

□ **COMMON SHELDUCK** *Tadorna tadorna*
Prior to 1974 the Common Shelduck was a rare summer visitor (which had bred occasionally from 1921) and scarce passage migrant in Nottinghamshire. Since then the number of breeding pairs and visiting birds has increased considerably - as in other parts of the English Midlands - as

birds have taken advantage of the worked out gravel pits along the Trent and Idle Valleys and the larger lakes of the Dukeries. More recently, breeding pairs have also colonised former colliery sites.

Common Shelduck - annual breeding pairs in Nottinghamshire 1975-2004

Years	1975-79	1980-84	1985-89	1990-94	1995-99	2000-04
Maximum pairs	12	26-29	12-14	16-22	34	30+

Each year numbers build up in the county between February and May. Most birds then disperse after breeding has taken place to traditional moulting sites although some remain in the winter months. Lound is a key site for this species with 18 pairs present in 2001 and 56 birds on 12.05.1995 and 13.03.2005 - the maximum counts for any one site 1974-2007.

☐ MANDARIN DUCK *Aix galericulata*

The first published record of Mandarin Duck for Nottinghamshire was an escaped bird at Attenborough 12.11.1963. However since 1977 the species has been an annual visitor to the county (except in 1979 and 1981-82) in small numbers.

Mandarin Duck - birds recorded in Nottinghamshire 1975-2005*

Years	1975-79	1980-84	1985-89	1990-94	1995-99	2000-04	2005	Total
Birds	5	5	5	14	c57	c135	c8	c229

obvious escapes are excluded from the above table.

No doubt many of the records involve escapes from wildfowl collections or released birds but some may originate from the large feral populations elsewhere in the country (for example in nearby Derbyshire).

Mandarin Duck are most regularly seen at Clumber Park and other sites in the Dukeries where a few birds were deliberately released in the 1970s and where there are pools surrounded by old trees and evergreen shrubs which meet the core requirements of this species. Breeding was proved in the Dukeries in 2000 and single pairs bred at Clumber Park, Welbeck and Sturton-le-Steeple in 2006 and at Clumber Park and Vicar Water in 2007.

Outside the breeding season, there have been 2 large counts from Clumber Park in recent years (15 on 01.01.2003 and 23 on 21.11.2004).

☐ EURASIAN WIGEON *Anas penelope*

The Eurasian Wigeon is an increasingly common winter visitor to Nottinghamshire and rare breeding bird. Whitaker mentioned a single 19th century breeding record involving a wounded ♀ at Moorgreen. No wild birds bred again until 1997 when 1 pair successfully bred at Lound. However between 2000 and 2004 1-5 pairs bred annually at 2 or more sites in the county and 3 pairs bred at Lound in 2007.

Numbers of wintering birds have increased steadily from around 800 in the early 1970s to an estimated 8000 in the county in February 2003. Site counts have shown a similar increase with 1056 at Cottam on 31.12.1993, then 1380 there on 16.01.1994, 2200 at Holme Pierrepont in February 1999 and a record 2609 at the same site in January 2006.

☐ AMERICAN WIGEON *Anas americana*

There are 8 county records of this North American vagrant. The first record was a ♀ at Netherfield 27.10-02.11.1969.
Another 7 birds have been found since 1974:

1986   ♂ Lound area 06.03.1986 & 05-10.01.1987 & 04-15.03.1987.
1998   1st winter ♂ Idle Stop, Lound and Hallcroft 08.03-27.04.1998 (with gaps).
2002   Eclipse ♂ Holme Pierrepont 22-25.09.2002.
       1st winter ♂ Hoveringham 28.09-16.10.2002.
       1st winter ♂ Erewash Valley and Bennerley Marsh 24.10 - 14.11.2002.
2004   1st winter ♂ Cotham Flash 05.11.2004.
2007   ♂ Hoveringham 01.01-17.03.2007.

See NBAR 2002 pp. 120-122.

☐ GADWALL *Anas strepera*

The Gadwall is an increasing species in Nottinghamshire - both as a breeding bird and as a winter visitor. Gadwall first bred in the county at Thoresby in 1968. By 2000 the county total was estimated to be possibly 70 pairs and 13 pairs bred at Lound alone in 1997 and subsequent years. Breeding numbers in Nottinghamshire are nationally significant as no more than 1000 pairs breed in Britain most years.

Wintering numbers have increased in the same period as shown in the table below:

Gadwall - maximum site counts in Nottinghamshire 1975-2007

Years	1975-79	1980-84	1985-89	1990-94	1995-99	2000-04	2005-07
Site maxima	28	41	160	162	413	398	486

The winter population is now over 600 birds with key concentrations in the Trent Valley (for example at Attenborough NR and Holme Pierrepont), the Dukeries (Thoresby and Carburton) and at Lound where a record 486 birds were counted on 28.12.2005.

☐ EURASIAN TEAL *Anas crecca*

The Eurasian Teal is an extremely rare breeding species and reasonably common winter visitor to Nottinghamshire. At present there is little suitable breeding habitat for this species in the county and Eurasian Teal bred less than annually in the period 1974-2007 with a peak of 6-7 pairs in 1980.

As a winter visitor small numbers are present on many undisturbed waters but larger concentrations sometimes occur as set out below:

Eurasian Teal - maximum site counts in Nottinghamshire 1975-2007

Years	1975-79	1980-84	1985-89	1990-94	1995-99	2000-04	2005-07
Site maxima	700	600	1000	770	1050	1120	1500

The most in the period 1974-2007 is a count of 1500 at Cottam Power Station on 06.02.2007.

☐ GREEN-WINGED TEAL *Anas carolinensis*
The North American counterpart of the Eurasian Teal has visited the Trent and Idle Valleys on 7 occasions since 1989. All records have involved ♂ birds.
1989   Cottam 04.03.1989.
1990   Lound 01-09.04.1990.
1995   Lound 11.01.1995.
1996   Attenborough NR 27.02.1996.
1997   Lound 07.04.1997.
1999   Holme Pierrepont 14.03-02.04.1999.
2004   A returning bird at Langford Lowfields 03-18.01.2004, 20.11.2004-12.02.2005, 23.10-18.12.2005, 05-26.11.2006 and 13.01.2007. Also seen at Girton 28.02.2005 and 31.01.2006 and at Collingham 21-22.12.2007.
See NBAR 1989 p.38.

☐ MALLARD *Anas platyrhynchos*
A common resident and winter visitor. It was estimated that 5-6000 wintered in the county in the early 1970s. Wintering numbers seem to have fallen a little in recent years as elsewhere in Britain. This contrasts with the fortunes of most other duck species and may be a result of increased habitat competition from species such as Gadwall. However the Mallard is still a common and familiar bird throughout the county.

Mallard - maximum site counts in Nottinghamshire 1975-2007

Years	1975-79	1980-84	1985-89	1990-94	1995-99	2000-04	2005-07
Site maxima	1481	1791	1200	1000	1235	970	986

Attenborough NR is a key site for this species with the largest count for 1974-2007 of 1791 in January 1980 at a time when over 7000 birds were in Nottinghamshire.

☐ PINTAIL *Anas acuta*
The Pintail is an uncommon passage migrant and winter visitor to Nottinghamshire. Counts of over 30 birds are unusual but there were 96 at Pump House Pool on the River Idle on 24.02.2003 in what was a very good year for this species. There have only been 5 other counts of 50 or more

birds in the county since 1974 - 50 west over Burton Meadows 12.09.2001, 50 at Newington 03.02.2003, 50 in the Idle Valley 14.02.2003, 60 at Girton 23.02.2003 and 52 at Langford Lowfields in December 2007.

Pintail - maximum annual site counts in Nottinghamshire 1992-2007

Years	1992	1993	1994	1995	1996	1997	1998	1999
Site maxima	23	9	29	25	11	16	22	26
Years	2000	2001	2002	2003	2004	2005	2006	2007
Site maxima	20	50	16	96	28	24	6	52

There are occasional reports of birds in the summer months and a pair probably bred in north Nottinghamshire in 1969.

□ GARGANEY *Anas querquedula*
A rare summer visitor and uncommon passage migrant through Nottinghamshire. Breeding was first proved in the Trent Valley at Nottingham SF in 1945 but there are few published records for 1974-2007 with certain breeding only established in 1976 (2 pairs), 1991 (1 pair), 2001 (2 pairs) and 2004 (1 pair). Attempts at breeding in the county appear to have peaked in the 1970s as set out below:

Garganey - breeding status in Nottinghamshire 1940 - 2007

Years	1940-49	1950-59	1960-69	1970-79	1980-89	1990-99	2000-07
Years bred	2	5	1	3	0	1	2
Years suspected	1	2	3	4	3	1	3

On migration, an average of 10-15 birds are recorded each year, as shown by records of non-breeding birds since 1996:

Garganey - migrant birds in Nottinghamshire 1996-2007

Years	96	97	98	99	00	01	02	03	04	05	06	07
Birds Feb-June	11	5	11	7	8	3	8	9	9	8	8	11
Birds July-Nov	10	7	7	11	11	13	9	3	6	5	8	10

Extreme dates 24.02.1952 (locality not recorded) - 20.11.1977 (Besthorpe).

□ BLUE-WINGED TEAL *Anas discors*
The only fully acceptable county record of this North American duck is of a ♂ at Lound on 26-27.04.2000, a bird which was also seen in East Yorkshire.
An earlier ♂ at Carburton Lane near Clumber Park on 28-29.01.1979 was probably an escape from captivity.
See NBAR 2000 pp. 139-140.

□ SHOVELER *Anas clypeata*
The Shoveler has been a rare breeding bird in Nottinghamshire since the nineteenth century. The species first bred at Rainworth Water in 1874.

Numbers peaked in the 1940s and 1950s when 30-45 pairs bred in the Nottingham area.

In the period 1974-2007 breeding numbers have been much lower with no more than 16 pairs in the whole of the county in the best year (2001). In contrast the species has increased as a non-breeding bird with peak numbers in late autumn. Attenborough NR was a key site until the early 1990s and the maximum count there was of 286 birds in October 1990. More recently the north of the county has become the key area for this species, with Clumber and Lound the most important sites. 300 were counted at Lound on 29.10.2005 - the best site count for the period 1974-2007.

☐ RED-CRESTED POCHARD *Netta rufina*

Historically this species was a rare visitor to Nottinghamshire with just 21 birds recorded between 1949 (when the first for the county was at Netherfield on 15.02.1949) and 1973. Birds were found in all months except January and July.

Between 1974 and 2007 Red-crested Pochards were recorded virtually annually and the pattern of occurrence of this species has been clouded by increasing numbers of escaped birds. Nevertheless records show a distinct peak between August and November which suggests that some birds may have been genuine visitors from continental Europe - or wanderers from feral populations elsewhere in England - rather than escapes from captivity.

Red-crested Pochard - birds recorded in Nottinghamshire 1975-1999*

Years	1975-79	1980-84	1985-89	1990-94	1995-99	Total
Birds	13	5	11	30	19	78

Months	J	F	M	A	M	J	J	A	S	O	N	D
Birds	3	1	0	2	3	2	5	21	12	5	15	9

Known escapes are excluded from the above table.

From around 2001 a small feral breeding population has become established at Lound with a maximum of 38 birds there on 15.10.2007 and small numbers have also been recorded in the Trent Valley in recent years.

☐ COMMON POCHARD *Aythya ferina*

This species is a common wintering bird in Nottinghamshire and the population has increased considerably since the early 1970s. A count of 754 at Attenborough NR on 22.12.1974 was at the time a record for the county but 1705 were at Misson on 27.02.1980 and a new record was set when 1760 were recorded at Hoveringham on 14.11.1995. Smaller numbers have been recorded in milder winters in recent years.

Breeding numbers have also increased. Common Pochard first bred in 1945 at King's Mill Reservoir and breeding was irregular for several years thereafter. However 1-15 pairs bred in 28 out of the 34 years 1974-2007 and breeding has been annual since 1988.

☐ **REDHEAD** *Aythya americana*
The first British and sole Nottinghamshire record of this American vagrant is of a ♂ at Bleasby and Gibsmere 08-27.03.1996.
Only one has been found in England since - a ♂ (perhaps the same bird) at Rutland Water, Rutland / Leicestershire 04-24.02.1997.
See *NBAR 1996 pp.89-91 and Birding World Vol. 9/3 (March 1996) pp.93-97.*

☐ **RING-NECKED DUCK** *Aythya collaris*
Nine adult or first winter ♂s of this North American duck have been recorded in Nottinghamshire since the first county record in 1983.
1983   The Dukeries (Carburton, Thoresby, Welbeck and Clumber) 24.04-04.06.1983.
1984   Carburton 30.09.1984.
1988   South Muskham 14-15.06.1988.
1990   Clumber 26.04-14.05.1990.
2000   Hoveringham and South Muskham 02.01-22.02.2000 (with gaps).
2001   Bleasby and Gibsmere 07.01-17.02.2001 (with gaps).
       Hoveringham 13-21.04.2001.
       Lound 27.10.2001
2004   Attenborough NR 15-29.10.2004.
See *NBAR 1988 p.41.*

☐ **FERRUGINOUS DUCK** *Aythya nyroca*
A very rare visitor to Nottinghamshire with just 18 birds recorded in the county between the mid 19th century and 1964 including 1♂ and 2♀ together at King's Mill Reservoir 06.02.1949 and 2 birds together on 4 other occasions. The next record was a ♀ at King's Mill Reservoir on 27.11.1971 and there have been just 1 or 2 records since 1974:
1981   A ♂ at South Muskham 12.12.1981 (which may also have visited Lound 25.10-29.11.1981 although the record was never published).
2002   An immature ♀ at King's Mill Reservoir 17.11.2002. This bird had a metal ring and was not assessed by the BBRC.
Extreme dates 29.08.1964 (Holme Pierrepont and Attenborough NR) - April1951(Attenborough NR).

☐ **TUFTED DUCK** *Aythya fuligula*
The Tufted Duck is another species which has benefited from the flooding of worked out gravel pits in the Trent Valley and elsewhere. Breeding was first recorded as long ago as the 1830s or 1840s at Osberton and became more regular in the 1950s. This trend continued 1974-2007 and 158 breeding pairs were reported in the county in 2000 with a minimum of 83 pairs at the Lound gravel complex. Wintering numbers have similarly

increased since the 1970s with large numbers in the Trent Valley in particular.

Tufted Duck - maximum site counts in Nottinghamshire 1975-2007

Years	1975-79	1980-84	1985-89	1990-94	1995-99	2000-04	2005-07
Site maxima	251	400	1044	1801	1500	928	798

South Muskham is a favoured site and the first four-figure count for the county was of 1000 there on 14.02.1985. Besthorpe - Girton is another popular site for this species and a record 1801 birds were there on 17.02.1991.

☐ **GREATER SCAUP** *Aythya marila*
This sea duck is an uncommon passage migrant and winter visitor to Nottinghamshire with an average of 10 to 20 birds recorded annually 1974-2007. 1991 was easily the best recent year for this species with 30 to 50 birds in the county. Most records fall between September and May but there are also occasional summer records between June and August. Double figure counts are exceptional with no flock between 1974 and 2007 exceeding the 12 at Netherfield 24.03.1996.
The county record is of 23 together at Besthorpe 30.11.1968.

☐ **LESSER SCAUP** *Aythya affinis*
There are 3 recent spring records of this increasing North American vagrant which was first recorded in Britain in 1987:
1990   ♂ Lound 22-23.04.1990 (the second English record).
1996   ♂ King's Mill Reservoir 15.04.1996.
1998   ♂ Rampton Lakes, Cottam 14-17.05.1998.
See NBAR 1990 pp.55-56.

☐ **COMMON EIDER** *Somateria mollissima*
This sea duck is a rare visitor to the county. The first record was a ♀ shot on floods near Nottingham 16.11.1882. The next record was surprisingly of up to 18 birds together at Attenborough NR 24.12.1972 - 17.02.1973 (2 of which wandered to Bulcote 31.12.1972).
Subsequent records are listed below:
1986   A moribund ♂ (with a Dutch ring) at Brinsley 08.02.1986.
1989   An immature ♂ at Colwick CP 05.12.1989.
1993   A second larger influx occurred in 1993 with 35-42 birds in the county 30.10-02.11.1993. Most were in the Trent Valley with 6-7 different birds at Gunthorpe 30-31.10.1993, up to 28 (7 ♂s and 21 immature/♀s) at South Muskham 31.10-01.11.1993, 1 at Hoveringham 31.10.1993, 1 at Colwick CP 31.10-01.11.1993 and 2 at Attenborough NR 02.11.1993. Singles were also at Wollaton

Park 31.10-01.11.1993, King's Mill Reservoir 02.11.1993 and in Retford 02.11.1993.
A later ♂ went through Colwick CP 13.11.1993.
1995   A ♀ or immature bird at Attenborough NR 10-27.12.1995.
2006   An immature at Langold Lake 16.10.2006.
The 1993 birds were part of a wider inland invasion which brought approximately 200 birds to Yorkshire, the East Midlands and the West Midlands.
See NBAR 1993 pp.93-94.

☐ LONG-TAILED DUCK *Clangula hyemalis*
The Long-tailed Duck is another wintering coastal species which rarely straggles inland. It has always been a very scarce visitor to Nottinghamshire with just 5 birds found in the 19th century and approximately 31 more birds between 1920 and 1973. All records before 1974 were found between October and February save for a ♂ at Wollaton Park 18.04.1954. All but 4 of these early records came from the Trent Valley with an influx of at least 10 birds there in February 1947 and 6 together at Netherfield 10.11.1957.
A maximum of 39 more birds were found between 1974 and 2007:

Long-tailed Duck - birds recorded in Nottinghamshire 1975-2007

Years	1975-79	1980-84	1985-89	1990-94	1995-99	2000-04	2005-07	Total
Birds	8	6	6	7	4	5	3	39

Months	J	F	M	A	M	J	J	A	S	O	N	D
Birds	5	1	1	1	1	2	1	0	1	6	9	11

31 of the 39 birds were recorded between October and January as might be expected from a winter visitor to Britain but this species is plainly less weather dependent than some other species with a number of unseasonal records into early summer. Several records involve lingering birds with a particularly long-staying bird at various localities along the River Trent 11.12.1976-01.05.1977 and another at Wollaton Park 18.12.1988-06.05.1989. Predictably many birds were recorded at sites in the Trent Valley but the Lound gravel pit complex has had 6 birds since 1982.

☐ COMMON SCOTER *Melanitta nigra*
This sea duck is primarily an uncommon but annual passage migrant through Nottinghamshire with a concentration of spring and autumn records (particularly in the Trent Valley) which suggests there is a regular overland migration through the county.

Common Scoter - birds recorded in Nottinghamshire 1975-2007

Years	1975-79	1980-84	1985-89	1990-94	1995-99	2000-04	2005-07	Total
Birds	129	56	54	393	186	157	100	1075

Months	J	F	M	A	M	J	J	A	S	O	N	D
Birds	29	8	29	169	86	103	23	15	75	260	251	27

There have been 2 exceptional flocks in the county during this period - 115 at Lound 22.10.1990 and 120 at Colwick CP 04.11.1994.

▢ VELVET SCOTER *Melanitta fusca*

A very rare visitor to the county. There were approximately 11 birds found in Nottinghamshire between 1884 (when a ♀ was shot at Welbeck 06.11.1884) and 1947 (when perhaps as many as 6 birds were in the Trent Valley between January and March).

2 further birds were recorded before 1974 - a ♂ at Besthorpe 24.11.1957 and an immature/♀ at Colwick CP 31.07.1973 (the only early record outside the period 6th November - 9th March).

Only 7 more birds were recorded between 1974 and 2007:

1976	A ♀ at Attenborough NR 11.02.1976.
1991	2 ♀ at South Muskham 12.11.1991.
1994	1 at Cottam 22.05.1994.
1995	A 1st winter ♂ at King's Mill Reservoir 14.03.1995.
	A ♀ at Girton 04-12(or15).11.1995.
1996	A ♂ at King's Mill Reservoir 16.11.1996.

▢ BUFFLEHEAD *Bucephala albeola*

A widely twitched drake at Colwick CP 17-26.03.1994 is the only county record. This was the 4th British record of this North American vagrant.
See NBAR 1994 pp.89-90 and Birding World vol. 7/3 (March 1994) pp.102-104.

▢ COMMON GOLDENEYE *Bucephala clangula*

An increasingly common winter visitor to the county. A significant influx of 80 birds was recorded in the hard winter of 1947 but that figure has been exceeded regularly since 1976. Monthly wildfowl counts for Nottinghamshire exceeded 100 for the first time in January 1982, 200 in February 1985, 300 in January 1987 and 400 in March 1996 with a peak of 489 birds in March 1997. The gravel pits of the Trent Valley (for example Colwick CP, Holme Pierrepont and Hoveringham) are particularly popular with this species and a record 200 were at Hoveringham on 04.01.2003.

Common Goldeneye - maximum site counts in Nottinghamshire 1975-2007

Years	1975-79	1980-84	1985-89	1990-94	1995-99	2000-04	2005-07
Site maxima	24	66	130	116	166	200	116

Numbers depart rapidly in the spring but up to 5 birds have summered annually in recent years.

☐ **SMEW** *Mergellus albellus*
An uncommon winter visitor with most records falling between late November and early March. Wintering numbers have increased since the mid 1990s and the last blank year was as long ago as 1983.

Smew - annual totals for Nottinghamshire 1975-2004

Annual total (birds)		0	1-10	11-20	21-50	50+
Number of years	1975-84	2	6	1	1	0
	1985-94	0	6	2	2	0
	1995-04	0	0	2	6	2

Many birds are found in the Trent Valley with large weather related influxes in some years. Record numbers were recorded in January 1997 with 70-100 birds in the county including up to 13 different birds at Attenborough NR and 12 at Hoveringham-Gibsmere. However there is an earlier count of 27 together in the Trent Valley during the hard winter of 1947. Birds are rarer elsewhere in the county but there have been virtually annual records in the Idle Valley in recent years.
Extreme dates 23.10.1999 (Lound) - 19.04.1997 (Girton) although a ♀ was apparently seen at Attenborough NR 21.06.1969.
See NBAR 1987 pp.39-43.

☐ **RED-BREASTED MERGANSER** *Mergus serrator*
This species is a scarce and irregular winter visitor to Nottinghamshire with annual records since 1984. Following a single bird at Attenborough NR on 01.12.1974 a minimum of 294 further birds were recorded up to 2007.

Red-breasted Merganser - birds recorded in Nottinghamshire 1975-2007

Years	1975-79	1980-84	1985-89	1990-94	1995-99	2000-04	2005-07	Total
Birds	72	35	30	48	70	26	13	294

Months	J	F	M	A	M	J	J	A	S	O	N	D
Birds	59	79	18	19	14	0	0	0	2	25	23	55

Occurrences of this species are often linked to hard winter weather and - as with several rare grebes and divers - there was a record influx of at least 53 birds between 16.02.1979 and 05.03.1979 with 19 birds at Winthorpe 24.02.1979. Over 400 birds were found inland in England in February 1979. The only other double figure count since 1974 is of 11 birds at Gunthorpe on 27.12.1980.
See NBAR 1987 pp.39-43.

□ GOOSANDER *Mergus merganser*
The Goosander is another species which has increased considerably as a winter visitor, particularly in the Trent Valley (for example at Besthorpe and Girton). Monthly county wildfowl counts have peaked at 184 (February 1979), 237 (January 1987), 313 (December 1990), 503 (February 1996) and there was an astonishing hard weather influx of 901 birds in January 1997. However since 1997 milder winters have seen far fewer Goosander in Nottinghamshire and the highest total for any one site remains the 160 birds counted at Besthorpe / Girton on 19.01.1985.
Odd birds have summered since the 1970s but there has been no evidence of local breeding in the county so far.
See NBAR 1987 pp.39-43.

□ RUDDY DUCK *Oxyura jamaicensis*
This introduced North American duck was first recorded in Nottinghamshire at Attenborough NR 08-17.03.1962. The Ruddy Duck has increased rapidly in the county and in the rest of England since the 1970s. The first breeding record for Nottinghamshire was in 1980 and the species bred at 16 sites in 1999. Two years later 22 pairs were counted at Lound.
Wintering and post-breeding numbers have shown a similar increase:

Ruddy Duck - maximum site counts in Nottinghamshire 1975-2007

Years	1975-79	1980-84	1985-89	1990-94	1995-99	2000-04	2005-07
Site maxima	6	44	119	131	165	220	202

The first three-figure site count involved 119 at Attenborough NR on 08.03.1986. By 2001, 4 sites (Clumber, Colwick CP, Lound and Holme Pierrepont) were recording three-figure counts of this species and 220 were counted at an unnamed site in January 2004. However the UK Government is currently undertaking a controversial nationwide cull of Ruddy Ducks which may eliminate this species in the future.

☐ RED GROUSE *Lagopus lagopus*

This species breeds in Derbyshire and Yorkshire and has very occasionally wandered to Nottinghamshire, often as a result of hard winter weather. There are six records for the county between 1860 and 1941 - 1 shot at Bevercotes in November 1860, several killed near Nottingham in the winter of 1863, 1 shot at Clipstone 04.01.1883, 1 at Attenborough in January 1888, 1 near Ratcher Hill in January 1903 and a small covey at Osberton in late autumn 1941.

There have only been 3 records since then:
1979   2 at King's Mill Reservoir 20.03.1979.
1986   A ♀ found dead near Gorseybrecks 20.03.1986.
1988   1 on the Bestwood estate 18.02.1988.

☐ BLACK GROUSE *Tetrao tetrix*

The Black Grouse is a lost species in Nottinghamshire, as it is in most of the rest of lowland England, a victim of changing agricultural practices and land use pressures.

In the mid to late nineteenth century, the species was a scarce breeding species in the county (for example at Newstead, Mansfield Forest, Sherwood Forest and Stapleford Wood) and 16 were killed in 1 day at Mansfield in approximately 1850. For Whitaker writing in 1907, the species was just hanging on in the county around Mansfield, Rufford, Budby and Warsop but there were only 3 subsequent records - a ♀ shot at Inkersall 11.12.1910, a ♂ at Ratcher Hill December 1914-January 1915 and a ♀ near Rainworth in autumn 1917.

Black Grouse are today found in just 5 of 28 English counties in which the species formerly bred and further records in Nottinghamshire seem to be highly unlikely.

☐ RED-LEGGED PARTRIDGE *Alectoris rufa*

A reasonably common introduced gamebird which was first recorded at Thoresby in October 1851. It is difficult to estimate the true size of the population of this species in the county as game releases are still common and large coveys often comprise recently released birds. With that significant qualification covey sizes in recent years at least indicate a steady county population as set out below:

Red-legged Partridge - maximum coveys in Nottinghamshire 1975-2007

Years	1975-79	1980-84	1985-89	1990-94	1995-99	2000-04	2005-07
Maximum annual coveys	15-29	19-70	16-66	20-80	19-82	21-100	28-38

Several of the recent counts involve sporting birds released at Oldcotes in 1999-2000 but 70 in several small coveys at Bunny Moor 18.11.1984 and 80 near Shelton 11.11.1990 may have involved 'wild' breeding birds. Odd birds also penetrate urban areas with - for example - 2 records from London Road (in 1998 and 2004) and one in the Park Estate (in 1978), near to the centre of Nottingham.

☐ **GREY PARTRIDGE** *Perdix perdix*
The Grey Partridge was formerly a common farmland bird with 752 brace shot in Nottinghamshire on 10.10.1906. The species has been in decline since the 1950s but it is still a well distributed bird in the county with records from 80 sites in 2000 and breeding season records from 49 sites in 2005. The Idle Valley is a stronghold for the species with at least 50 breeding pairs recorded in 1994 and the only 3 figure count for the county 1974-2007 - 144 birds (in 5 coveys) on 09.10.1993. However birds are still released by shooting syndicates (for example 1400 in the Idle Valley in 1995) which distorts the true status of this species.

☐ **COMMON QUAIL** *Coturnix coturnix*
The Common Quail is a scarce summer visitor to Nottinghamshire with numbers varying greatly from year to year. In the period 1974-2007 there were 5 blank years (the last in 1991) and at least 16 other years in which less than 10 adults were recorded. Conversely there were several good 'Quail years' in this period:

Common Quail - large influxes into Nottinghamshire 1974-2007

Years	Breeding information
1992	c19 calling birds.
1995	38-39 calling birds.
1997	c30 breeding pairs.
1998	c66 calling birds.
2001	29 calling birds.

In most years, the Idle Valley is the stronghold for this species (particularly around Gringley and Misson) with at least 48 calling in the best year 1998. However in good years other parts of the county hold small numbers of calling birds.
Extreme dates 18.04.2002 (Gringley Carr) - 10.11.1982 (Newstead Abbey). In addition Joseph Whitaker makes reference to an old December record of a bird shot at Ramsdale and there are several other old claims for December and January. The only recent winter record is of an escaped bird in gardens at Arnold 15-16.12.1990.

☐ **COMMON PHEASANT** *Phasianus colchicus*
A common but exceptionally poorly recorded resident species in the county. Joseph Whitaker records that over 2000 birds were once killed in one day

at Welbeck but attempts to assess the current population of this species are hampered by the large numbers of birds which are released each year. The largest group recorded 1974-2007 - 450 at West Leake hills 07.11.1998 - was made up of recently released birds.

☐ RED-THROATED DIVER *Gavia stellata*

A rare visitor to the county since the nineteenth century when 6 birds were recorded. A further 18 were found in Nottinghamshire between 1947 and 1972. All were recorded between early November and early May except for 1 at Besthorpe 15.05-12.06.1960.

Following a bird at Hoveringham 16-17.02.1974 (which was found dead on 16.03.1974) another 23 were recorded in the county to 2007. Over half of the birds recorded (12) were part of a hard winter influx between January and April 1979.

Red-throated Diver - birds recorded in Nottinghamshire 1975-2007

Years	1975-79	1980-84	1985-89	1990-94	1995-99	2000-04	2005-07	Total
Birds	12	2	1	2	1	3	2	23

Months	J	F	M	A	M	J	J	A	S	O	N	D
Birds	7	8	1	2	0	0	0	0	0	1	0	4

Most records since 1974 are from the Trent Valley with 5 records from Colwick CP in that period. As with other hard winter visitors, this species has become a real county rarity in recent years with only 5 birds since 1996 - at Colwick CP 28.12.2000, Gunthorpe and Hoveringham 23.10-04.11.2001, at Gunthorpe and Hoveringham again 04-08.12.2002, at Balderton 17-25.01.2006 and at Welbeck 01-15.04.2007.

☐ BLACK-THROATED DIVER *Gavia arctica*

The Black-throated Diver is a rare visitor to Nottinghamshire with just one record in the nineteenth century - a bird near Worksop in January 1848. A further 8 birds were found between 1947 and 1973 with all records between November and April.

Another 20-21 birds have been found since 1974 including 5-6 in the Trent Valley during hard winter weather in February 1979. Only 2 recent records (at Wollaton Park 11.02.1990 and the 1996 Bestwood bird discussed below) have involved birds at sites away from the Trent Valley.

Black-throated Diver - birds recorded in Nottinghamshire 1975-2007

Years	1975-79	1980-84	1985-89	1990-94	1995-99	2000-04	2005-07	Total
Birds	6-7	1	9	1	1	1	1	20-21

Months	J	F	M	A	M	J	J	A	S	O	N	D
Birds	2	9-10	5	1	0	0	0	0	0	0	1	2

This species has become a real county rarity due to the recent run of mild winters with only 3 records since 1990 - a long staying bird at Bestwood 08.03.1996 - 05.06.1996 (when it was captured and moved to Daneshill Lakes where it remained to 09.06.1996), 1 at Attenborough NR 27.01-01.02.2003 and 1 at Hoveringham 09-13.12.2007.
Extreme dates 04.11.1973 (Hoveringham) - 25.07.1976 (Hoveringham).
See NBAR 1996 pp. 104-105.

☐ **GREAT NORTHERN DIVER** *Gavia immer*
There are only 23 records of this species in Nottinghamshire and it is the scarcest of the three diver species to visit the county. There are 2 records from the 19th century, 3 1946-59 and another in 1970, with all dated records between December and February.
Between 1974 and 2007 a further 17 were recorded:

Great Northern Diver - birds recorded in Nottinghamshire 1975-2007

Years	1975-79	1980-84	1985-89	1990-94	1995-99	2000-04	2005-07	Total
Birds	3	0	3	3	3	3	2	17

Months	J	F	M	A	M	J	J	A	S	O	N	D
Birds	2	0	1	1	0	0	0	0	0	1	7	5

Many of these birds have stayed for several days, notably a bird at King's Mill Reservoir, Bleak Hills Pond and Daneshill 23.12.1989-22.02.1990 (with gaps) and another at Hoveringham 07.12.2005-31.01.2006. Of the 17 recent records, 10 birds were found in the Trent Valley (as were 5 of the 6 earlier birds) with 3 at Colwick CP. Away from the River Trent, 4 birds were at King's Mill Reservoir and 2 at Lound.
Extreme dates 28.10.2002 (King's Mill Reservoir) - 22.04.1996 (Colwick CP).

☐ **LITTLE GREBE** *Tachybaptus ruficollis*
A reasonably common resident species in Nottinghamshire. It was estimated that up to 80 pairs were breeding in the county in the early 1970s and there may be many more today as several flooded gravel pits have been created since then.
From 1974 until the early 1990s King's Mill Reservoir was the most important site in the county with 8-15 pairs breeding and a peak count of 117 birds (January-February 1979). In more recent times Holme Pierrepont has been the best county site for this species (and one of the top 5 sites in England) with up to 30-35 pairs (1999) and a record count of 156 birds on 11.09.1999. In the north of the county Lound is a key site for this species with 37 breeding pairs in 2000 and a peak of 109 birds on 16.09.2001.

☐ **GREAT CRESTED GREBE** *Podiceps cristatus*
This species was widely persecuted for its feathers in nineteenth century Britain. Since then it has benefited from increased protection and - as with Little Grebe - from the creation of flooded gravel lakes. Census counts show the increase of this species in Nottinghamshire with just 27 pairs in 1932, 110-120 adult birds (1949), 101-110 (1955), 233-236 (1965) and an estimate of 376 adult birds and 131 breeding pairs in 1975. There are probably even more birds today as new pits have been created since 1975. Two Trent Valley sites - Attenborough NR and South Muskham - have been particularly significant for this species in the last 30 years. At Attenborough NR 30-40 pairs were reported in 1989 with a county record count of 228 birds on 02.02.2006 and at South Muskham 133 birds were counted on 18.12.1988.

☐ **RED-NECKED GREBE** *Podiceps grisegena*
A very scarce and irregular winter visitor to Nottinghamshire, the rarest of the 5 grebes to be recorded in the county. There were 6 19th century records and a further 25 birds between 1929 and 1970 including 9 in February-March 1947.
There were no further records between 1971 and 1977 but birds were recorded in 20 of the next 30 years.

Red-necked Grebe - birds recorded in Nottinghamshire 1975-2007

Years	1975-79	1980-84	1985-89	1990-94	1995-99	2000-04	2005-07	Total
Birds	23-25	0	17	3	13-16	13	5	74-79

Months	J	F	M	A	M	J	J	A	S	O	N	D
Birds	8-10	31-33	5	1	4	0	1	1	3	7	7	6-7

Occurrences of this species are influenced by hard weather (as in 1947) and an unprecedented 21-23 birds were found during the hard winter of December 1978- March 1979. Approximately 480 birds were found in Britain and Ireland during that winter.
There is one summer arrival during the period 1974-2007 - a bird at Gunthorpe 01.07.1988 - which follows earlier records of a bird shot on the River Trent at Clifton in June 1850 and another bird at Netherfield on 02.06.1968. An injured bird also lingered at Misson until 04.06.1979.

☐ **SLAVONIAN GREBE** *Podiceps auritus*
A scarce winter visitor and passage migrant through Nottinghamshire. Only 26 birds were found in the county between 1838 and 1973 but these included a hard weather influx of 8 birds in January-February 1947.
Birds were less than annual in the period 1974-2007 with records in only 25 out of the 34 years. However the Slavonian Grebe has increased considerably since 1993 with records in all years except 2007, as summarised below:

Slavonian Grebe - birds recorded in Nottinghamshire 1975-2007

Years	1975-79	1980-84	1985-89	1990-94	1995-99	2000-04	2005-07	Total
Birds	14	2	8	11	22-27	15	5	77-82

Months	J	F	M	A	M	J	J	A	S	O	N	D
Birds	12-13	18-20	7-8	5	0	0	0	0	2	11	11	11-12

9 birds were found in the county 22.01-28.02.1979 during a period of hard winter weather. By contrast there are no summer (May-August) records for the county.

Many records come from the Trent Valley with 13 birds at Hoveringham and 10 at Holme Pierrepont since 1974. Elsewhere in the county 7 were found at King's Mill Reservoir between 1974 and 2007.

□ **BLACK-NECKED GREBE** *Podiceps nigricollis*

This is another grebe which has increased in Nottinghamshire in the last 30 years. Traditionally this species was a rare winter visitor and passage migrant with only 3 birds reported between 1974 and 1978. Since then birds have been reported in all years except 1984.

Following one at Attenborough NR 17.01.1974 another 194 birds were recorded in Nottinghamshire (excluding known breeding birds) up to 2007 as recorded below:

Black-necked Grebe - non-breeding birds in Nottinghamshire 1975-2007

Years	1975-79	1980-84	1985-89	1990-94	1995-99	2000-04	2005-07	Total
Birds	2	15	33	45	55	29	15	194

Months	J	F	M	A	M	J	J	A	S	O	N	D
Birds	8	3	12	24	59	4	10	23	19	17	13	2

There is inevitably some duplication of records in the above table as birds have moved around in the county. However it is clear from the monthly distribution of records that the species is primarily an uncommon passage migrant through the county rather than a true wintering bird and it is significant that Black-necked Grebes were not caught up in the influx of other rare grebes and divers into the county in 1979.

Black-necked Grebe bred for the first time in Nottinghamshire in 1979 and another pair bred at the same site in 1985. A third pair at this site was robbed in 1993 but since 1997 1 to 9 pairs have bred or attempted to breed at up to 3 sites per year. However egg collecting continues to limit breeding success.

□ **FULMAR** *Fulmarus glacialis*

A rare visitor to Nottinghamshire which was first recorded in the county at Colwick Sluices on 4th June 1971. There have been 25 county records since then with all records falling between 1975 and 1997, except for a bird at Lound on 11.06.2006:

Fulmar - birds recorded in Nottinghamshire 1975-2007

Years	1975-79	1980-84	1985-89	1990-94	1995-99	2000-04	2005-07	Total
Birds	2	2	9	5	6	0	1	25

Months	J	F	M	A	M	J	J	A	S	O	N	D
Birds	1	0	1	0	3	4	2	6	6	1	0	1

The spread of records suggests that the occasional appearances of this species are less linked to storms at sea than is the case with other rare seabird species. Many have occurred at well watched localities in the Trent Valley with 3 records at Colwick CP and at Attenborough NR and 4 at Holme Pierrepont since 1974. In addition an unprecedented 6 birds were seen together at Plumtree on 02.08.1988.

☐ **MANX SHEARWATER** *Puffinus puffinus*

The Manx Shearwater is a rare visitor to Nottinghamshire with just 13 birds found between 1882 and 1967. All were found between 1st and 23rd September except for 1 at Attenborough NR 27.06.1952 and 1 at Bestwood in late August 1888.

There have only been 28 records since 1967:

Manx Shearwater - birds recorded in Nottinghamshire 1975-2007

Years	1975-79	1980-84	1985-89	1990-94	1995-99	2000-04	2005-07	Total
Birds	5	5	10	5	2	0	1	28

Months	J	F	M	A	M	J	J	A	S	O	N	D
Birds	0	0	0	0	1	1	0	4	22	0	0	0

25 birds were picked up either dead or exhausted and all but 3 birds were found between 28th August and 25th September. This strongly suggests that the majority are post-breeding birds displaced by autumn storms. The other 3 records in the period are of single birds at Attenborough NR 18-19.05.1979 (the only spring record for the county), at Ollerton 16.06.1997 and at Newark 07.08.1987.

☐ **EUROPEAN STORM PETREL** *Hydrobates pelagicus*

A very rare storm driven visitor to Nottinghamshire with 11 birds found prior to the First World War. Further single birds were caught alive at Carlton 26.10.1945, found dead at Staythorpe 31.10.1952 and at Stragglethorpe 01.11.1952 and one was caught by a cat at Kimberley on 17.11.1957.

Only 3 have been found since:
1977   1 at Hoveringham 17.09.1977.
           1 found dead at East Markham 16.10.1977.
1982   1 picked up exhausted at Aslockton 27.11.1982.

☐ LEACH'S STORM PETREL *Oceanodroma leucorhoa*
The Leach's Storm Petrel is a rare storm-driven visitor to the county. However this species is more prone to be blown inland than European Storm Petrel.
7 birds were found between September and December from 1840 to 1910. The species was next recorded in 1952 when a large wreck of Leach's Storm Petrels occurred in October and November with around 7000 birds found throughout Britain. 10 birds were found in Nottinghamshire between 29th October and 5th November 1952.
None were then discovered until 1977 but there have been 18 birds since with the majority of records in the Trent Valley.

Leach's Storm Petrel - birds recorded in Nottinghamshire 1975-2007

Years	1975-79	1980-84	1985-89	1990-94	1995-99	2000-04	2005-07	Total
Birds	4	2	7	3	1	1	0	18

Months	J	F	M	A	M	J	J	A	S	O	N	D
Birds	0	0	0	0	1	0	0	0	10	3	2	2

15 birds occurred between 3rd September and 19th November and another 2 were found in December - a pattern which is consistent with storm driven appearances inland. Several of these birds have been found either dead or dying including the only spring record for Nottinghamshire at Stanton-on-the-Wolds 08.05.1977.

☐ NORTHERN GANNET *Morus bassanus*
The Northern Gannet is a rare and irregular vagrant to Nottinghamshire.
Only 34 birds were found in the county between 1833 and 1972 including 4 at Daybrook 22.09.1953 and 4 at Carlton 16.04.1972.
An immature bird was then found at Aslockton 09.09.1974 and was followed by 21 others between 1975 and 2007. 18 of the 22 birds were recorded between 9th September and 24th November and several have been found dead or exhausted, suggesting that many birds are driven inland by stormy unsettled weather.

Northern Gannet - birds recorded in Nottinghamshire 1975-2007

Years	1975-79	1980-84	1985-89	1990-94	1995-99	2000-04	2005-07	Total
Birds	4	5	0	1	2	7	2	21

Months	J	F	M	A	M	J	J	A	S	O	N	D
Birds	1	0	1	0	2	0	0	0	6	7	4	0

Records are widely scattered through the county with records from a number of unusual localities such as Hodsock, Ruddington, Laneham and Cromwell.

☐ GREAT CORMORANT *Phalacrocorax carbo carbo*
The status of the Great Cormorant has been transformed in the last 30 years. In 1974 the Great Cormorant was a scarce inland visitor but it is now a commonly encountered species particularly in the Trent Valley. This species has no doubt benefited from the extensive gravel workings in the Trent Valley - many of which have been stocked by fishing clubs - but this is also part of a national trend with increasing numbers feeding and breeding inland.

Two sites are central to the success of this species. Attenborough NR has consistently held the largest groups of birds in the county with increasing winter site counts of 47 (February 1979), 55 (January 1985), 107 (December 1986), 160 (December 1987), 200 (December 1989), 290 (December 1990), 351 (December 1991), 441 (February 1996), and 459 (23.01.1997 - the county record). Numbers have levelled off or declined slightly in recent years but Attenborough NR remains a key site and nesting has been reported there since 2004. The other significant site is Besthorpe where the species bred for the first time in the county in 1990. This site holds the only large colony in the county with as many as 178 pairs in 1999:

Great Cormorant - breeding pairs in Nottinghamshire 1990-2007

Sites	1990	1991	1992	1993	1994	1995	1996
Besthorpe	9	33	74	92	114	140	176
Attenborough NR	0	0	0	0	0	0	0
Sites	1999	2000	2001	2003	2004	2006	2007
Besthorpe	178	98	78	120	100	90	137+
Attenborough NR	0	0	0	0	1	0	c20

The continental sub-species *P.c.sinesis* occurs regularly in Nottinghamshire and most breeding records probably involve this form. See NBAR 1990 p.65.

☐ SHAG *Phalacrocorax aristotelis*
This seabird is a very scarce autumn and winter visitor to Nottinghamshire with a handful of records (just 16 birds) between 1851 and 1972.
Since then birds have been recorded in 21 of the 34 years to 2007 as set out below:

Shag - birds recorded in Nottinghamshire 1975-2007

Years	1975-79	1980-84	1985-89	1990-94	1995-99	2000-04	2005-07	Total
Birds	5	5	7	74-93	9	8	9	117-136

Months	J	F	M	A	M	J	J	A	S	O	N	D
Birds	7	62-81	6	1	0	0	0	11	12	4	10	4

As can be seen from the above tables an exceptional influx occurred in February 1993 with 56 at Sutton Lawn's Park 01-03.02.1993 with numbers decreasing thereafter through to 19.02.1993. 19 other birds in the county between 05.02.1993 and 29.03.1993, probably involved dispersing birds from Sutton Lawn's Park. This exceptional event was probably the result of winter storms and other birds were found inland in Derbyshire, Leicestershire and the West Midlands region at the same time.

☐ EURASIAN BITTERN *Botaurus stellaris*

The Eurasian Bittern was historically a rare winter visitor to Nottinghamshire but small numbers have overwintered regularly in recent years. Attenborough NR has become the key site with records in 23 years 1974-2007 and 1-3 birds wintering annually since 1994, principally in the large Delta reedbed. All records at Attenborough NR fall between 15th August and 16th April except for a single bird on 16.07.1979.

Away from Attenborough NR, there were 69 published records from 28 sites 1974-2007 with 14 birds recorded at Lound, 12 at Holme Pierrepont and 8 at Hoveringham.

Eurasian Bittern - birds recorded in Nottinghamshire (excluding records from Attenborough NR) 1975-2007

Years	1975-79	1980-84	1985-89	1990-94	1995-99	2000-04	2005-07	Total
Birds	4	6	9	12	6	18	14	69

Months	J	F	M	A	M	J	J	A	S	O	N	D
Birds	16	14	5	2	1	0	0	3	3	3	9	13

All 69 records fall between 5th August and 11th April with the exception of an injured bird at Lound 12.05.1993-31.12.1993 which was eventually taken into care.

☐ LITTLE BITTERN *Ixobrychus minutus*

There were 4 records of this vagrant heron in Nottinghamshire up to 1921 - a ♂ Sutton Reservoir, Mansfield March 1858, a ♂ on the River Lin near Worksop 24.05.1870, 1 on the River Trent between Dunham and Marnham 21.08.1889 and a ♀ at Watnall 22.03.1921. Predictably all were shot.
No birds have occurred since then.

☐ NIGHT HERON *Nycticorax nycticorax*

This southern European heron has visited Nottinghamshire 4 times. The first was a bird shot at South Clifton in autumn 1820. 148 years later an adult was at Ratcliffe-on-Soar 28.04.1968. Finally there were 2 records in the 1970s:

1976   A bird at Langley Mill Flashes in the Erewash Valley 09.09-02.10.1976 was reported to have flown into Nottinghamshire on 09.09 & 02.10.1976 (although the BBRC Annual Report for that year makes no mention of this).

1977   An immature bird was at Colwick Hall Lake 12.08.1977.

In addition an escaped bird was recorded at Attenborough NR intermittently between 2000 and 2004.

◻ **SQUACCO HERON** *Ardeola ralloides*
There are 3 records of this European heron from Nottinghamshire. A ♂ was killed at Bestwood Park in August 1871 and an immature bird was seen at Nottingham SF on 27.08.1944.
Over fifty years later a third bird visited the county:
1998   An adult bird at Attenborough NR on 27.06.1998 moved to nearby Martin's Pond 29.06-01.07.1998.

◻ **CATTLE EGRET** *Bubulcus ibis*
A bird at Radcliffe-on-Trent on 04.04.1999 was the only county record of this rare but increasing southern European vagrant prior to 2008.
See NBAR 1999 p.131.

◻ **LITTLE EGRET** *Egretta garzetta*
This south European egret has colonised Britain in the last 20 years and now breeds in several southern and eastern English counties.
The first for Nottinghamshire was one on the River Trent near Littleborough 29.05.1970 and another was at Misson 09-19.08.1981. However since 1993 Little Egrets have been recorded in 13 out of the 15 years to 2007 with records of at least 32 birds at 9 sites in 2005 and as many as 19 birds present in the Collingham / Langford Lowfields area on 25.09.2007.

Little Egret - birds recorded in Nottinghamshire 1975-2007

Years	1975-79	1980-84	1985-89	1990-94	1995-99	2000-04	2005-07	Total
Birds	0	1	0	1	c12	c69	c105	c188

The majority of birds have occurred between May and September but small numbers have overwintered in recent years.
Most Little Egret records have come from the Trent Valley where the key sites are Attenborough NR, Holme Pierrepont, Netherfield, Gunthorpe/ Burton Meadows, Hoveringham and Langford Lowfields/Collingham New Workings. Lound has also attracted up to 10 birds in recent years with birds dispersing to feed along the River Idle.
The first confirmed breeding record may not be far away as copulating birds were observed in the Trent Valley at Barton-in-Fabis in 2005.

◻ **GREAT WHITE EGRET** *Ardea alba*
There are 5 county records of this European egret. The first was a bird killed at Osberton some time before 1838 (c.1835) - the second or third British record. Another 4 have been found since 1974:
1989   1 at Attenborough NR 16-17.06.1989 (which was also seen in Cornwall and Gloucestershire).
2003   1 at Holme Pierrepont 20.07.2003 and at Attenborough NR 21.07.2003.
2007   1 at Lound and Newington intermittently 03-25.09.2007.
        1 at Attenborough NR 21.10.2007.

□ **GREY HERON** *Ardea cinerea*
The Grey Heron is a reasonably common resident species in Nottinghamshire. The breeding population has fluctuated somewhat with 88-90 nests in 5 recorded heronries in a 1928 census. However following the loss of a large colony at East Stoke (peaking at 40-50 pairs) through persecution in the early 1950s numbers were much reduced and only 17 pairs were recorded as breeding in the county in 1965. Since then the species has recovered and there were as many as 236 pairs nesting in 1999. 28 heronries were used between 1950 and 2000 and the largest current colonies are set out below:

Grey Heron - large heronries in Nottinghamshire still in use in 2007

Heronry	Breeding information
Brandshill Wood	Founded 1974. Peak 70 nests in 1997. Nesting pairs relocated to Attenborough NR in 2007.
Osberton	Founded 1959. Peak 66 nests in 1999.
Besthorpe	Founded 1971. Peak 51 nests in 1997.

However 3 other large colonies have been abandoned in the last 50 years, perhaps as a result of disturbance:

Grey Heron - large heronries in Nottinghamshire lost since 1958

Heronry	Breeding information
Colwick	Peak 66 nests 1952. Last bred 1958 (save for 1 pair in 1998).
Clumber	Peak 53 nests 1954. Last bred 1959.
Welbeck	Peak 54 nests 1995. Last bred 1999. Breeding attempt 2007.

Large numbers of birds are sometimes counted at the end of the breeding season and there were a succession of large counts at Holme Pierrepont from the 1970s to the early 1990s with a maximum of 128 birds counted there 15.09.1991.

□ **PURPLE HERON** *Ardea purpurea*
This southern European heron has been found in the county on 7 occasions. There is one nineteenth century record of a bird shot on the River Trent at Clifton, Nottingham in 1868. There were no records for the next 100 years but another was found at Carburton Lakes 18.03.1972.
There are 5 subsequent spring and early summer records for the county:
1978   1 at Carburton Lakes 06.06.1978.
2000   1 at King's Mill Reservoir 29.05.2000.
2001   1 at Gringley Carr 22.04.2001.
2003   1 at Attenborough NR 26.04-02.05.2003.
           1 at Colwick CP 27.04.2003.
See NBAR 2003 p.111

□ **WHITE STORK** *Ciconia ciconia*
The White Stork was recorded twice in 19th century Nottinghamshire - 1 was shot near Bawtry in 1825 and 2 were killed from a larger flock on the

River Trent near to the same locality in 1829. Between 1829 and 1973 there was one further record of a single bird between Pilgrim Oak and Hutt House 12.04.1915.

17 have occurred since then and records have been virtually annual since 1996 with several records no doubt relating to escaped birds:

White Stork - birds recorded in Nottinghamshire 1975-2007

Years	1975-79	1980-84	1985-89	1990-94	1995-99	2000-04	2005-07	Total
Birds	1	0	0	0	5	7	4	17

Months	J	F	M	A	M	J	J	A	S	O	N	D
Birds	0	0	0	5	3	4	2	1	1	0	1	0

1977 1 at Bradmore and East Leake 08-09.05.1977.
1996 1 at Welbeck RW 08.06.1996.
1998 1 at Clumber, Shireoaks and Blyth 27.04.1998.
      Another at Holme Pierrepoint, Colwick CP and Attenborough NR 27.04-01.05.1998.
1999 1 at Clumber, Gamston, West Bridgford, Attenborough NR and Welbeck 01.05-17.05.1999.
      Another at Oxton 10.07.1999.
2000 1 at Carburton, near Oxton, King's Mill Reservoir and Hallcroft 06.04-23.04.2000.
2001 1 at Clumber 28.04.2001.
      Another at West Stockwith 11.05.2001.
2002 1 at East Stoke 12.06.2002.
2003 1 at Gedling 05.06.2003.
2004 1 at Welbeck RW 05.06.2004.
      1 at Hoveringham 05.07.2004.
2005 1 at Lound, Barnby Moor and Retford 23.08-01.09.2005.
      1 at Bassingfield, Holme Pierrepont and Nottingham city centre 14.11.2005.
2006 1 near Retford 03-07.09.2006.
2007 1 at Lound 18.04.2007.

☐ BLACK STORK *Ciconia nigra*

This striking European vagrant has reached Nottinghamshire on 3 occasions. In the nineteenth century one was apparently shot at Colwick in autumn 1871 (though several authors have doubted this record and no specimen apparently still exists).

Two more have occurred in recent years:
1999 An elusive bird in the Vale of Belvoir at Cropwell Bishop, Colston Basset and Kinoulton 13 & 16.06.1999.
2002 A bird over Blidworth Pit Top 02.06.2002.

See NBAR 1999 pp.129-130.

☐ **GLOSSY IBIS** *Plegadis falcinellus*
There is at least one old record of this species - a bird shot at Misson in north Nottinghamshire on 04.10.1909. 14 birds were recorded in Yorkshire during the same month. Another bird is said to have been at this site on 18.11.1857. No birds have occurred since.

☐ **EURASIAN SPOONBILL** *Platalea leucorodia*
The Eurasian Spoonbill is a very rare visitor to Nottinghamshire - as it is to other inland counties - with 6 birds between 1831 and 1952 and 15 records of 17 birds since.

Eurasian Spoonbill - birds recorded in Nottinghamshire 1975-2007

Years	1975-79	1980-84	1985-89	1990-94	1995-99	2000-04	2005-07	Total
Birds	1	2	4	2	3	2	3	17

Months	J	F	M	A	M	J	J	A	S	O	N	D
Birds	0	0	0	1	5	1	3	4	2	0	1	0

Lound is the favoured locality for this species with 4 one-day May records since 1992 and 8 birds were recorded at 7 different sites in the Trent Valley between 1976 and 2007 including 3 birds at Langford Lowfields.
Extreme dates 20.04.1989 (Ollerton) - 06.11.1987 (Stapleford).
In addition there are two undated 'winter' records from the 19th century.

☐ **HONEY-BUZZARD** *Pernis apivorus*
For many years the Honey-Buzzard was the jewel in the crown of Nottinghamshire's breeding birds with 1-2 well-known pairs breeding in the Dukeries in most years from the 1960s to 2001. This is a rare breeding bird in England with published nesting records from just 15 counties and many birders made the pilgrimage to Welbeck Raptor Watchpoint to see these elusive birds. Successful breeding has not been confirmed since 2001 but birds have continued to return to the site.
Migrants are occasionally seen elsewhere in the county. However in the autumn of 2000 there was an unprecedented national influx of c500 Honey-Buzzards which brought 68 birds to Nottinghamshire 21.09-04.10.2000 with 7 together over King's Mill Reservoir (23.09.2000) and 7 over Kirkby-in-Ashfield (29.09.2000) the best counts.
Extreme dates 12.05.1997 (Osberton) - 05.10.2003 (Hoveringham).

☐ **BLACK KITE** *Milvus migrans*
1 found dead at Dewhurst Plantation, Osberton c. 27.05.1978 is the only county record of this European vagrant. It had apparently been poisoned. Many British records involve overshooting migrants in April and May.
See NBAR 1985 pp.54-56.

☐ **RED KITE** *Milvus milvus*
This species probably bred in Nottinghamshire in the early 19th century. However between 1876 and 1973 only 3 birds were seen in the county, one in 1947 and two in 1972. In recent years there has been a considerable increase in records linked to the successful reintroduction schemes in Northamptonshire/Leicestershire, Yorkshire and elsewhere.

Red Kite - birds recorded in Nottinghamshire 1975-2005

Years	1975-79	1980-84	1985-89	1990-94	1995-99	2000-04	2005	Total
Birds	1	3	0	2-4	15	c40	9	c70-72

Months	J	F	M	A	M	J	J	A	S	O	N	D
Birds	1-3	3	10	15	6	11	1	8	5	3	3	4

A record 29 birds were found in 2006 with 7-8 more in 2007 and the first Nottinghamshire breeding record for two centuries may not be far away.

☐ **WHITE-TAILED EAGLE** *Haliaeetus albicilla*
There are 3 old records of this huge raptor from Nottinghamshire - 1 spent three weeks at Welbeck Park in the winter of 1838, 1 was present 05-08.11.1896 (when it was shot) at Park Farm, Annesley and another bird suffered a similar fate at Cropwell Butler in November 1916.
One has occurred since:
2007   1 over Budby Common 31.03.2007.

☐ **MARSH HARRIER** *Circus aeruginosus*
The Marsh Harrier was only recorded 9 times in Nottinghamshire before 1974 with 1 killed at Thoresby Park in 1848 and 8 birds 1945-72. It has increased considerably since then with one further bird in 1974 and annual records from 1977.

Marsh Harrier - birds recorded in Nottinghamshire 1975-2004

Years	1975-79	1980-84	1985-89	1990-94	1995-99	2000-04	Total
Birds	7	11-13	c19	c35	c98	c114	284-286

The species is now a regular non-breeding summer visitor (particularly to the Idle Valley) with many records involving immature birds or failed breeding birds from the Humberhead Levels. There are also a few winter records in Nottinghamshire.
The first breeding record may not be far away - in August 2004, 2 adults and 2 juveniles were found in the Idle Valley. It was thought that they may have bred over the border in South Yorkshire.

☐ **HEN HARRIER** *Circus cyaneus*
The Hen Harrier is a regular but uncommon winter visitor to Nottinghamshire where numbers may have decreased slightly from the early 1990s. Most records come from the Idle Valley (particularly Gringley Carr) which is a significant staging area for this species. Close studies of

this population have demonstrated that a number of individual birds move through each winter (eg. a minimum of 11 different birds January-April 1996 and 16 birds January-May and October-December 1998).
Records are rare away from the Idle Valley. As an example there are few recent records from the many well-watched sites in the Trent Valley:

Hen Harrier - Birds found in the Trent Valley, Nottinghamshire 1985-2007

Years	1985-89	1990-94	1995-99	2000-04	2005-07	Total
Birds	1	4	2	3	3	13

There are no breeding records for Nottinghamshire but a ♀ or possibly a pair summered in the Dukeries in 1976.

☐ **MONTAGU'S HARRIER** *Circus pygargus*

The Montagu's Harrier has always been a rare visitor to Nottinghamshire although a pair summered at Whitwell, Derbyshire in 1955 and 1956 and were seen in Nottinghamshire. Breeding may have taken place locally. However only 8 other birds were recorded between 1884 and 1969.
There were 15 further records between 1974 and 2007 with all dated records falling between 29th April and 4th September.

Montagu's Harrier - birds recorded in Nottinghamshire 1975-2007

Years	1975-79	1980-84	1985-89	1990-94	1995-99	2000-04	2005-07	Total
Birds	1	1	1	1	5	5	1	15

Months	J	F	M	A	M	J	J	A	S	O	N	D	Undated
Birds	0	0	0	1	4	3	5	1	0	0	0	0	1

Most birds were brief migrants and 10 of the 15 records involved one-day birds.

1978	1♂ at Treswell 25.06.1978.
1984	1 at Welbeck in the spring.
1986	1♂ at Rainworth 19.07.1986.
1994	1st summer ♀ at Misterton Carr 10.07.1994.
1995	1♀ at Misterton, Gringley and Everton Carrs 29.04.1995.
	1♂ at Idle Stop 30.05.1995.
1996	1♂ near Warsop 29.07.1996.
1997	1♀ in the Idle Valley 27.07-03.08.1997.
1999	1♂ on the Osberton Estate 04.05.1999.
2000	1♀ at Langford Lowfields 08.05.2000.
	1 immature at Creswell Crags, Welbeck and Shireoaks 16-21.08.2000.
2001	1♂ at Papplewick Moor 19.06.2001.
	1 immature ♂ at Gringley Carr 01.07-04.09.2001.
2003	1♀ at West Stockwith 06-08.05.2003.
2006	1 at Holme Pierrepont 24.06.2006.

Extreme dates 02.04.1884 (near Ratcher Hill) - 23.10.1945 (Upton Fields).

☐ **NORTHERN GOSHAWK** *Accipiter gentilis*
The Northern Goshawk was an extremely rare visitor to Nottinghamshire until the late 1960s. The only records were 1 shot at Rufford in 1848, an intriguing record of a displaying pair near Oxton 11.04.1945 and a juvenile near Worksop 26.12.1947.
In the late 1960s and early 1970s there were a few reports of escaped falconers birds and deliberate releases which were followed by records of between 8 and 14 birds 1974-1987. However from 1988 a small population has gradually built up in the county with 1-6 pairs either proven or suspected of breeding in the Dukeries and around Birklands in most years 1998-2007.
Welbeck RW is often a good place to observe this species in early spring.

☐ **EURASIAN SPARROWHAWK** *Accipiter nisus*
This species underwent a catastrophic decline in the early 1960s caused primarily by the introduction of toxic food dressings into the food chain. As a result the Eurasian Sparrowhawk had become a scarce resident in the county in the 1960s and early 1970s. However the species recovered in the 1970s, 1980s and into the 1990s and is now widespread throughout the county in a range of woodland, suburban and urban habitats.

☐ **COMMON BUZZARD** *Buteo buteo*
Historically the Common Buzzard was a scarce visitor to Nottinghamshire which bred in the 1820s and was a scarce resident from 1935 to 1958 without breeding being proved.
Between 1974 and 1988 the Common Buzzard remained a scarce visitor with less than 20 records per year. However from at least 1989 birds were present in the Dukeries throughout the year and the first recent record of breeding in Nottinghamshire was published in the Nottinghamshire Birdwatchers Annual Report for 1991, although breeding had been suspected in earlier years. By 1996 there were several pairs in the Dukeries and 2 pairs in the south of the county. 4 years later resident birds were present at 33 sites and this increase has continued through to 2007 when there were over 500 reports in the county including 30 counted at Welbeck 15.03.2007. The current strongholds for this species are the woods of the Dukeries and the patchwork of woodland in the Trent Valley to the east and west of Nottingham. Further increases and consolidation can be expected.

☐ **ROUGH LEGGED BUZZARD** *Buteo lagopus*
The Rough Legged Buzzard is a rare wintering bird in England away from the East Coast but birds were found reasonably regularly in Nottinghamshire up to the First World War and over 30 were said to have been killed in Nottinghamshire and Leicestershire in 1839-40. However only 1 bird was found between 1915 and 1975 - a bird shot at Annesley 25.10.1926.

Since 1974 just 15 birds have been found in the county as set out below:

Rough Legged Buzzard – birds recorded in Nottinghamshire 1975-2007

Years	1975-79	1980-84	1985-89	1990-94	1995-99	2000-04	2005-07	Total
Birds	4	1	3	1	3	1	2	15

Months	J	F	M	A	M	J	J	A	S	O	N	D
Birds	2	4	2	1	0	0	0	0	0	1	4	1

Records of this species are well scattered but there are 3 records from King's Mill Reservoir since 1994. 12 of the 15 birds since 1974 were only seen on one day but a popular bird lingered at Oxton Dumble (having first been seen at nearby Epperstone) between 13.12.1998 and 18.03.1999.

1975   1 picked up at Clifton 18.02.1975 which died later.
1976   1 at Clumber 22.11.1976.
1979   1 at Littleborough 06.01.1979.
       1 at Walesby 10-17.02.1979.
1984   1 at Moorgreen 23.03.1984.
1985   1 at Headon 02.11.1985.
1986   1 at Girton 27.02.1986.
1988   1 at Budby 06-10.11.1988.
1994   1 at King's Mill Reservoir 23.10.1994.
1995   An immature at King's Mill Reservoir 19.02.1995.
1996   1 at Hallcroft 22.01.1996.
1998   1 near Epperstone and at Oxton Dumble 13.12.1998-18.03.1999.
2003   1 at King's Mill Reservoir 30.03.2003.
2006   1 at Gringley Carr 28.04.2006.
2007   1 at Lound 03.11.2007.

☐ OSPREY *Pandion haliaetus*

Originally a rare passage migrant, the Osprey is another bird which has shown a steady increase since the mid 1960s as the species has re-established itself in Britain.

Birds were recorded annually from 1974 and 1994 as set out below:

Osprey – birds recorded in Nottinghamshire 1975-1994

Years	1975-79	1980-84	1985-89	1990-94	Total
Birds	15	18	24	46	103

Months	J	F	M	A	M	J	J	A	S	O	N	D	Undated
Birds	0	0	1	25	40	12	3	8	9	3	0	0	2

Since 1994 numbers have increased further with 10-20 birds through the county in all years save 2002, 2003 and 2007 when less than 10 birds were recorded.

1-4 birds have summered in the Dukeries (particularly at Welbeck Great Lake) on several occasions since the mid 1980s and breeding could occur there in the future.

Extreme dates 25.03.2006 (Misterton Carr and Welbeck RW) - 16.11.1889 (locality unrecorded).

☐ COMMON KESTREL *Falco tinnunculus*
The Common Kestrel is a reasonably common resident breeding species throughout Nottinghamshire. A 3 year survey from 1979 to 1981 recorded an estimated 219 territories in the county at a density of around 1 pair per 10 square km. However reports for 2005-2007 indicate a recent decline in the population in Nottinghamshire.
See NBAR 1981 pp.32-35.

☐ RED-FOOTED FALCON *Falco vespertinus*
There are 4 records of this vagrant falcon in Nottinghamshire. The first county record was a 1st summer ♂ at Colwick CP 31.05-06.06.1973, arriving as part of a large national influx.
Since 1974, 3 more have been found:
1990   A ♀ in the Erewash Valley and at Langley Mill Flashes, Derbyshire on 16-17.07.1990 was apparently first seen just over the border in Nottinghamshire (although the BBRC report for that year makes no mention of the fact).
2000   An adult ♂ at Lound 12.05.2000.
2002   A 1st summer ♀ at Gringley Carr 12-17.06.2002.
May and June are the peak months for vagrants to Britain.

☐ MERLIN *Falco columbarius*
The Merlin is an uncommon passage migrant and winter visitor to Nottinghamshire. Recorded sightings increased in the 1990s with the recognition that the Idle Valley was an important staging and wintering area for this species. As an example 7-9 birds were seen in the Idle Valley between January and May 1996 and at least 16 passed through in August and September of the same year. However, there may have been a decline in the number of birds wintering in Nottinghamshire in the last decade.
There is a markedly northern bias to records of this species within the county and it remains a very scarce bird in the rest of Nottinghamshire with most other records coming from well watched sites in the Trent Valley.
According to Whitaker one or two pairs of Merlin nested in Sherwood Forest at Inkersal in the 19th century but there have been no breeding records since then.

☐ HOBBY *Falco subbuteo*
Prior to 1974 this species was a rare summer visitor to Nottinghamshire which had been suspected of breeding in the county. However, it has increased significantly since c1965 with a large number of sightings throughout the county each summer.

Breeding was first proved in 1979 (although Whitaker and Sterland suggested that the Hobby bred in Nottinghamshire in the nineteenth century). There has been a steady build up in breeding numbers since 1979 to a peak of at least 13 pairs in 1999.

Extreme dates 06.04.1999 (Wollaton Park) - 31.10.1982 (locality not recorded). Whitaker and Sterland refer to several November and December records in the mid nineteenth century (for example at Clifton in December 1849 and Newstead Park 09.12.1877) but these records are not now generally accepted. A record of one at Strelley 03.03.1849 may similarly be unreliable.

☐ PEREGRINE FALCON *Falco peregrinus*
Prior to 1974 the Peregrine Falcon was a rare and irregular visitor to Nottinghamshire, particularly after the national crash in the population of this species from 1956 onwards. Only 16 birds were found in the county between 1956 and 1973.

This trend continued between 1974 and 1989 with just 27 records:

Peregrine Falcon - birds recorded in Nottinghamshire 1975-1989

Months	J	F	M	A	M	J	J	A	S	O	N	D
Birds	4	2	4	1	1	0	1	1	4	1	2	6

Since 1990 the number of records has increased considerably in line with the recovery of the national population of this striking falcon, with 80 sightings from 32 localities in 1998 and 135 reports from 50 sites in 2005.

Pairs have summered on a number of buildings (for example at Ratcliffe-on-Soar Power Station and the Newton Building, Nottingham Trent University) since the late 1990s and breeding was proved for the first time in the county when 2 pairs bred in quarries in 2004. In 2005 4 pairs bred and a total of 6 pairs were present in the county in 2006 and 2007. Breeding birds have reoccupied several other midlands counties in the last 15 years and further breeding records may be expected.

☐ WATER RAIL *Rallus aquaticus*
A very scarce and unobtrusive breeding bird in Nottinghamshire. Breeding was first proved in the north of the county in 1971 and small numbers have been suspected of breeding in most years since 1984 with a maximum of 9 possible pairs in 2004.

Water Rail - total proven breeding records for Nottinghamshire 1975-2007

Years	1975-79	1980-84	1985-89	1990-94	1995-99	2000-04	2005-07
Breeding pairs	0	1	4	0	1	15	0

Otherwise the Water Rail is an uncommon winter visitor (principally between November and March) to thick reedbeds throughout the county. Small numbers winter at many of the mature gravel pits along the Trent Valley, particularly at Attenborough NR where 15 were ringed in November-

December 1996 and where as many as 30 were believed to be present in November 1998.

Water Rail - passage and wintering birds in Nottinghamshire 2000-2005

Years	January - April		August- December	
	Sites	Birds	Sites	Birds
2000	21	34	23	44
2001	20	42	24	51
2002	24	54	24	51
2003	27	57	16	25
2004	20	30	16	33
2005	23	43	16	48

□ **SPOTTED CRAKE** *Porzana porzana*

This species has always been a rare passage migrant in Nottinghamshire - particularly in the autumn - with 15 birds seen between 1944 and 1973.
Just 18 or 19 more birds were recorded between 1974 and 2007 and only 7 birds have been found since 1983:

Spotted Crake - birds recorded in Nottinghamshire 1975-2007

Years	1975-79	1980-84	1985-89	1990-94	1995-99	2000-04	2005-07	Total
Birds	9-10	2	2	0	2	2	1	18-19

Months	J	F	M	A	M	J	J	A	S	O	N	D
Birds	1	1	0	1	1	1	0	4	2	2	5-6	0

1976 was by far and away the best year with 6 or 7 birds recorded. Only 5 birds have stayed more than one day, the longest a confiding bird at Eakring Flash 16.08-15.09.2001.
Attenborough NR (9 or 10 birds) and Hoveringham (2 birds) were the only sites with more than one record in this period.
Joseph Whitaker reports that this species bred in the Nottingham marshes in or around 1871. However breeding has not been recorded in the county since that time.

□ **SORA** *Porzana carolina*

A ♂ at Attenborough NR 12.12.2004 to 01.01.2005 was the first county record of this North American crake and the thirteenth to be recorded in Britain.
The only previous record for an inland county is of a bird shot near Newbury, Berkshire in October 1864.
See NBAR 2004 pp116-117 and Birding World vol. 17/12 (December 2004) pp. 500-501.

□ **LITTLE CRAKE** *Porzana parva*
There are 4 records of this nationally rare vagrant. The first was a ♂ trapped at West Burton on 28.09.1970. This bird was followed by a remarkable run of records from the reedbeds at Attenborough NR:
1975   1 Attenborough NR 03.01.1975.
1976   1 Attenborough NR 06-c27.11.1976.
1983   1 Attenborough NR 05-20.11.1983.

□ **BAILLON'S CRAKE** *Porzana pusilla*
There are 3 old records of this vagrant crake which was a fairly regular visitor to this country in the nineteenth century - an immature bird shot at Bolam c.1891, an adult dead under wires at Gedling on 22.06.1893 and a bird seen at Rainworth on the unusual dates 02-03.12.1921, 05.03.1922 and 03.05.1922.

☐ **CORN CRAKE** *Crex crex*
This species was common in Nottinghamshire in the 19th century but thereafter declined rapidly in the county - as in other parts of Britain - as changes in agriculture took place. As a result the species was lost as a breeding bird around 1935. However odd pairs summered in the county into the 1940s and there were at least 2 further breeding attempts between 1950 and 1970 (1-2 summering at Burton Joyce in 1956 and a nest with 10 eggs accidentally destroyed in the Trent Valley north of Newark in 1968). 2 pairs may also have bred at Blidworth in 1965.
Since 1974 the species has been an extremely rare visitor to Nottinghamshire with only 7 records:
1975   3 calling at Misterton 29.06.1975.
1978   1 calling at Newstead Abbey 22.04.1978 onwards.
1984   1 calling at Gunthorpe 04.05.1984.
1993   1 at Sutton-in-Ashfield 05-06.09.1993 (when taken into care).
2001   1 at King's Mill Reservoir 03.10.2001.
2002   1 at Tithby 07.09.2002.
2006   1 near King's Mill Reservoir 08.05.2006.

☐ **MOORHEN** *Gallinula chloropus*
The Moorhen is a common and adaptable resident species exploiting a wide range of large and small waters throughout Nottinghamshire. As an example, 40 territories were recorded on the Chesterfield Canal between Worksop and Retford in 1981, 58 pairs bred at Lound in 2001 and 99 pairs bred at Centre Parcs in 2006.
Outside the breeding season many of the largest counts of this species have come from the north of the county (for example at Clumber Park and at Lound). However the highest Nottinghamshire count for 1974-2007 is of 213 birds at Lound 01.06.2007. An example of the adaptability of this species is a record of 89 birds feeding on fallen apples in an orchard at Worksop in late December 1992.

☐ **COMMON COOT** *Fulica atra*
The Common Coot is a common resident species which tends to prefer the larger waters within the county and has benefited from the creation of flooded gravel pits in the Trent Valley and elsewhere. Key sites include Hoveringham, Holme Pierrepont, Netherfield, Colwick CP and Girton in the Trent Valley and Clumber and Lound in north Nottinghamshire. A survey revealed a maximum of 287 breeding pairs in 1957 and approximately 470 pairs were estimated to be breeding in the county in 1977. Numbers are far higher today. For example only one pair was recorded at Sutton-cum-Lound in 1957 but 179 pairs attempted to breed at Lound gravel pits in 2001.
Winter numbers have shown a similar increase from perhaps 1000 birds in the 1950s with county counts of 3429 (January 1976), 4865 (January 1992) and 6170 (January 1997). Lound set the record count for the period 1974-

2007 with 1997 counted on 01.12.2006 and Holme Pierrepont had 1800 on 08.11.1996.
See NBAR 1977 pp.33-39.

□ **COMMON CRANE** *Grus grus*
There is one Victorian record of this species - an immature ♂ shot on the River Trent at Gunthorpe in January 1851.
There were no further records until 1979 but there have been 15 birds since:

Common Crane - birds recorded in Nottinghamshire 1975-2007

Years	1975-79	1980-84	1985-89	1990-94	1995-99	2000-04	2005-07	Total
Birds	1	0	0	1	4	5	4	15

Months	J	F	M	A	M	J	J	A	S	O	N	D
Birds	0	1	5	1	5	1	0	0	0	1	1	0

1979   1 in the Newington - Idle Valley area and at Lound 20.06.1979-13.09.1979 (with gaps).
1994   1 at Gringley Carr (and apparently also in flight over Lound) 18.02-04.03.1994 (with gaps).
1996   2 birds at Kersall and Norwell Woodhouse c01-09.03.1996.
1997   1 west over Clipstone Forest 28.05.1997.
1998   1 at Stoke Bardolph, Bulcote, Gunthorpe, Colwick CP, Holme Pierrepont, Netherfield and Hoveringham 18-30.10.1998.
2000   2 birds over Attenborough NR 21.03.2000 and what was presumed to be 1 of the same birds over Eakring 22.03.2000, Cotham Flash 01.04.2004, and at Tollerton, Holme Pierrepont and Stoke Bardolph 15.04.2000.
2001   1 at Besthorpe 11.04.2001.
         1 at Bole 20.05.2001.
2002   1 at Hoveringham 12.05.2002.
2005   1 at Bingham 03.11.2005 and then over Netherfield 05.11.2005.
2006   1 over Ravenshead 30.03.2006 and then over Annesley Pit Top 01.04.2006.
2007   1 at Attenborough NR 04.05.2007.
         1 at Misterton Carr 06.05.2007.

□ **LITTLE BUSTARD** *Tetrax tetrax*
There are two rather confusingly documented records of this species in Victorian Nottinghamshire - a bird killed at Shelton in the autumn of 1842 and another bird shot at South Clifton on 21.12.1866.
No birds have occurred since then.

☐ **GREAT BUSTARD** *Otis tarda*
The Great Bustard bred in England until the early nineteenth century. Surprisingly only one old record is generally accepted for Nottinghamshire - a bird at South Collingham in the Trent Valley on 01.04.1906 and 23-24.04.1906. This bird may have been a wanderer from a reintroduction attempt in Norfolk in 1900.

☐ **OYSTERCATCHER** *Haematopus ostralegus*
The Oystercatcher was historically a rare passage migrant through Nottinghamshire and one breeding attempt was recorded in 1969. However the species has colonised the county (and other parts of the Midlands) as a breeding bird since the first successful nesting record in 1979.

Oystercatcher - total breeding records in Nottinghamshire 1975-1994

Years	1975-79	1980-84	1985-89	1990-94
Breeding pairs	1	12	21-29	22-32

Since 1994 numbers have continued to increase with a minimum of 17 pairs attempting to breed in 2001 and breeding attempts at 15 sites in 2004 but failure rates are high. Many pairs use worked out gravel pits and Lound has been the key site since at least 1994. 9 pairs attempted to breed at Lound in 2001 and 26 birds were counted there on 14.07.2002.
The number of passage migrants has also increased slightly since 1974 and there are four large counts from Lound - 23 birds on 08.03.1993, 15.04.2002 and 31.07.2005 and 26 on 16.04.2006.

☐ **BLACK-WINGED STILT** *Himantopus himantopus*
There are 4 records of this southern European vagrant. The first was found at Perlethorpe on 30.01.1848. Almost one hundred years later 2 (or possibly 3) pairs bred at the old Nottingham SF 13.05-02.09.1945. 4 were also recorded at Nottingham SF on 05.05.1946, possibly returning birds from 1945. The 1945 breeding record was the first successful nesting record of this species in Britain and the event has been commemorated in the logo adopted by Nottinghamshire Birdwatchers. The only other successful breeding record of this species in Britain is of a pair at Holme, Norfolk in 1987.
Between 1974 and 2007 one further bird was recorded:
1974   1 at Attenborough NR 18.05.1974.

☐ **AVOCET** *Recurvirostra avosetta*
Historically this species was a rare visitor to Nottinghamshire but numbers passing through the county have increased recently in line with the growing breeding population along the River Humber and elsewhere in Britain. Just 30 birds were found in the county between 1800 and 1972. All were recorded in the Trent Valley except for 1 killed at Edwinstowe 24.07.1856 and these included 5 at Bulcote 13.05.1960 and 11 at Netherfield 22.05.1972.

Following 2 birds in May 1974 (singles at Attenborough NR and at Wilford) a further 63-64 were recorded 1975-2007 as set out below:

Avocet - birds recorded in Nottinghamshire 1975-2007

Years	1975-79	1980-84	1985-89	1990-94	1995-99	2000-04	2005-07	Total
Birds	13	2	0	6	7	15-16	20	63-64

Months	J	F	M	A	M	J	J	A	S	O	N	D
Birds	0	1	8-9	21	16	10	2	1	1	3	0	0

Only 9 of the 65-66 birds found since 1974 have stayed more than a day and none has stayed longer at a site than one at Besthorpe 23-26.06.1995. Since 1974 birds have been found at 18 different sites with 42-43 in the Trent Valley (including 14 at Hoveringham and 6 at Attenborough NR). Away from the River Trent, 18 birds have been found at Lound and Hallcroft Pits since 1997. The largest group in the period was 6 birds at Hoveringham on 08.04.1979.

□ **STONE-CURLEW** *Burhinus oedicnemus*

The Stone-Curlew is a lost breeding species in Nottinghamshire. It was a summer migrant to central parts of the county (for example at Papplewick, Oxton and Newstead) in Whitaker's day but last bred near Rainworth in 1891. The loss of Nottinghamshire's breeding birds was an early sign of a more general decline in the fortunes of this species in England. The Stone-Curlew also became extinct as a breeding bird in both Yorkshire and Lincolnshire in the first half of the 20th century. There are only 4 records in the last 100 years. The first was at Nottingham SF on 13.09.1953.

Since 1974, 3 more have been found:
1988   1 at Holme Pierrepont 09.04.1988.
1997   1 at Netherfield 23.05.1997.
2005   1 at Misterton Carr 24.04.2005.

See NBAR 1988 p.37 and NBAR 1997 pp.140-142.

□ **LITTLE RINGED PLOVER** *Charadrius dubius*

An uncommon breeding species and fairly common passage migrant which was first recorded in the county in the Trent Valley east of Nottingham in 1947. Breeding began in the Trent Valley in 1956 and the species has benefited from the creation of flooded gravel pits and the abandonment of some industrial sites since then. Two surveys in the period have revealed healthy numbers with 39 breeding pairs in 1984 and 52 pairs (33 confirmed breeding pairs) in 1999. In the later survey 32 pairs were using former colliery sites and 12 pairs were at gravel pit sites.

Small numbers are recorded on passage with 30 at Hoveringham 15.07.1993 the largest count for 1974-2007.

Extreme dates 10.02.1961 (Nottingham SF) - 09.11.2006 (Attenborough NR).

☐ **RINGED PLOVER** *Charadrius hiaticula*
A fairly common passage migrant which first bred successfully in the county in 1974 (having attempted to breed on 3 occasions since 1966). Breeding numbers then increased steadily to a peak of 22-23 pairs in 1984 and have levelled off at 8-17 pairs in recent years. Many breeding records have come from gravel pit sites but colliery pit tops are also used regularly.
The largest passage group reported 1974-2007 is 64 at Netherfield in May 2001.
Two subspecies have been recorded in the county *C. h. hiaticula* and the migrant **Tundra Ringed Plover** *C. h. tundrae*.

☐ **KILLDEER** *Charadrius vociferus*
The first and only Nottinghamshire record of this North American vagrant was a bird at Lound on 21.04.1981. There have been 50 British records.

☐ **KENTISH PLOVER** *Charadrius alexandrinus*
The Kentish Plover is a rare visitor to Britain which formerly bred on the South Coast. It is exceptionally rare inland and has reached Nottinghamshire just 11 times with 9 records before 1974 - near Bradder's Pond, Rainworth 13.04.1904, Nottingham SF 23.05.1947, 10-22.08.1947, up to 3 20-21.09.1947, 14.07.1950, 03.09.1952 and 18.04.1959, Attenborough NR 03.06.1963 and Holme Pierrepont 08.04.1966.
Since 1974 2 more birds have been found:
1979   1 at Bleasby 24.05.1979.
1997   1 at Gringley Carr 03.06.1997.

☐ **DOTTEREL** *Charadrius morinellus*
A reasonably regular spring migrant with records in the Idle Valley (and particularly at Gringley Carr) in 20 of the 34 years between 1974 and 2007. All records involve birds which were found between 6th April and 1st June except for a sickly bird 14.06-06.07.1975 and the largest trip was of up to 38 birds 27.04-24.05.1980:

Dotterel - spring trip sizes Idle Valley & Gringley Carr, Nottinghamshire 1974-2006

Years	74	75	76	77	78	79	80	81	82	83	84
Largest group	18	11	0	28	0	0	38	17	14	2	0
Years	85	86	87	88	89	90	91	92	93	94	95
Largest group	0	6	10	0	2	8	0	6	0	12	2
Years	96	97	98	99	00	01	02	03	04	05	06
Largest group	19	9	3	0	3	0	0	15	0	0	3

Only 3 records in the Idle Valley since 1974 have involved returning autumn birds with 6 in the Idle Valley 28.08.1994, 1 there on 05 & 13.09.1998 and 1 at Misson 08-09.09.1984.

Elsewhere in the county there were only 3 records between 1974 and 2007 with 2 at the former Wilford Power Station 16.05.1991 and singles at Netherfield 26.08.1995 and 23.05.1997. There were formerly more records away from the northern tip of the county including trips of c30 at Ruddington in April 1884 and c24 at Oxton Warren c1860.
See NBAR 1986 pp.40-43.

☐ AMERICAN GOLDEN PLOVER *Pluvialis dominica*
There are 6 autumn records of this species, all since 1984:
1984	1 at Idle Stop and Misson (and later in Humberside) 08-09.09.1984.
1995	1 at Hoveringham and Holme Pierrepont 10.05-12.05.1995.
	1 at Besthorpe and Girton 17.09-05.10.1995.
1998	1 at Gringley Carr and Lound 05-15.11.1998.
2000	1 at Sutton-on-Trent/Weston and Langford Lowfields 21.09 & 08.10.2000.
2001	1 at Gringley Carr 27.10-08.11.2001.

Most Nottinghamshire birds were found after careful searching of larger flocks of European Golden Plover. September and October are the peak months for birds to turn up in Britain.
See NBAR 1984 pp.39-40 and NBAR 1995 pp.110-113.

☐ EUROPEAN GOLDEN PLOVER *Pluvialis apricaria*
This species is a common passage migrant and winter visitor to Nottinghamshire with large flocks at traditional wintering sites such as Bennerley Marsh (maximum 5200 26.12.1998) Netherfield (maximum 6000 13.12.1992), Lound (maximum 8000 26.11.1992) and in the Idle Valley (maximum 14000 11-12.02.1995).

Four figure winter flocks are not uncommon in many other parts of the county and as many as 27,000 were estimated to be present in Nottinghamshire in the winter of 1994-95. However records are rare between mid May and July as birds return to their northern breeding grounds.

2 subspecies occur in the county - the **Southern Golden Plover** *P. a. apricaria* and the **Northern Golden Plover** *P. a. altifrons*.

☐ GREY PLOVER *Pluvialis squatarola*
The Grey Plover has always been an uncommon annual migrant through Nottinghamshire with peak numbers in May and September as shown by records during the thirteen years 1995-2007:

Grey Plover - birds recorded in Nottinghamshire 1995-2007
Months	J	F	M	A	M	J	J	A	S	O	N	D
Birds	7	5	7	7	59	3	1	9	55	33	18	6

Between 1974 and 1996 this species was a reasonably frequent visitor and there were 3 autumn double-figure counts in the early 1990s - 12 at

Besthorpe-Girton 21.09.1993, 13 at Lound 07.09.1995 and an exceptional 24 at the same site 11.09.1990. By contrast numbers have been fairly low since 1997 with no group exceeding 4 birds recorded in the county. The largest site count is of 28 birds at Colwick 11.01.1945.

☐ SOCIABLE LAPWING *Vanellus gregarius*
There is only 1 record of this vagrant in Nottinghamshire - a short stayed bird at Ranskill on 30.05.1978. There had been 41 British records by 2007.

☐ NORTHERN LAPWING *Vanellus vanellus*
The Northern Lapwing is still a reasonably common resident breeding bird in Nottinghamshire but has been in decline since the 1960s. Pairs now only breed in significant numbers in those areas where farming is not intensively managed but birds will use gravel pits, old pit tops and other bare wasteland sites.

Outside of the breeding season the Northern Lapwing is a familiar passage migrant and winter visitor in large numbers with many birds moving through in hard weather.

Northern Lapwing - maximum site counts in Nottinghamshire 1975-2007

Years	1975-79	1980-84	1985-89	1990-94	1995-99	2000-04	2005-07
Maximum site counts	4000	700	4000	7000	3100	3000	2000
Counts of 2000-7000 birds	3	0	11	25	16	8	1

The largest count for 1974-2007 is of 7000 at Holme Pierrepont on 17.12.1992.

☐ RED KNOT *Calidris canutus*
An uncommon passage migrant and scarce winter visitor. None at all were recorded in 1974 but records have been annual since then.

Red Knot - records for Nottinghamshire 1975-2007

Years	1975-79	1980-84	1985-89	1990-94	1995-99	2000-04	2005-07
Records	20	23	28	41	34	46	22
Maximum counts	9	9	13	11	51	7	4

Migration through Nottinghamshire is prolonged with a build up of records between March and May and again between August and November as shown by details of the 337 birds recorded from 1985 onwards:

Red Knot - birds recorded in Nottinghamshire 1985-2007

Months	J	F	M	A	M	J	J	A	S	O	N	D
Birds	32	4	15	65	47	4	12	40	53	30	26	9

During the period 1974-2007 there were only 4 double-figure counts for Nottinghamshire - 13 at Hoveringham 23.09.1985, 11 at Rainworth 23.10.1990, 10 at Clumber 04.01.1993 and an exceptional 51 birds at Lound on 18.04.1997.

☐ **SANDERLING** *Calidris alba*
An uncommon passage migrant with small numbers passing through the county between March and June in all but one year since 1974. The exception was 1982 when no birds were seen in the county at all. Even smaller numbers pass through in the autumn (July - October) but records are less than annual with 9 blank years from 1974. Winter records (November - February) are rarer still with birds in just 8 of the last 34 years. Two large movements have been recorded in recent years - 63 birds through the county including 45 at Lound on 15.05.1994 and 25 also at Lound on 13.05.2002. These are the only double-figure counts for Nottinghamshire since 1974.

☐ **LITTLE STINT** *Calidris minuta*
An uncommon annual migrant through Nottinghamshire. Numbers are relatively small during spring migration (April - June) with just 67 birds found in 19 years between 1975 and 2007 and a maximum of just 6 birds at any one site. By contrast the Little Stint is an annual autumn migrant (July - October) with double-figure site counts in several years and a peak of c115 birds through the county in autumn 1996. Odd birds linger into November and December but none overwintered in Nottinghamshire between 1974 and 2007.
The best count for the period is 26 birds at Girton 25.09.1996 but this is well down on the record count for the county - 44 birds at Nottingham SF 12.09.1946.

☐ **TEMMINCK'S STINT** *Calidris temminckii*
The Temminck's Stint has always been a scarce and irregular visitor to Nottinghamshire although there are many records for Nottingham SF between 1944 and 1963 (involving a minimum of 2 May birds and another 32 between July and October including a record count of 7 there 14.09.1946).
Following 1 in August and 2 in September 1974 a further 45 were recorded between 1975 and 2007 as set out below:

Temminck's Stint - birds recorded in Nottinghamshire 1975-2007

Years	1975-79	1980-84	1985-89	1990-94	1995-99	2000-04	2005-07	Total
Birds	6	6	5	4	5	15	4	45

Months	J	F	M	A	M	J	J	A	S	O	N	D
Birds	0	0	0	2	22	0	1	7	9	3	1	0

The majority of the 48 birds recorded since 1974 have been in the Trent Valley with 11 birds at Netherfield/Colwick, 9 at Hoveringham, 4 at Girton and 4 at Holme Pierrepont.
Extreme dates   spring 14.04.1995 (Hoveringham)-25.05.1997 (Lound).
                autumn 15.07.2000 (Cotham)-08.11.1981 (Hoveringham).

☐ **BAIRD'S SANDPIPER** *Calidris bairdii*
This North American wader is a recent addition to the county list with a single moulting adult at Lound on 19-24.08.1998.
See NBAR 1998 p.162.

☐ **PECTORAL SANDPIPER** *Calidris melanotos*
The Pectoral Sandpiper was recorded in Nottinghamshire on 4 occasions before 1974 - at Nottingham SF 26-30.09.1948 and 20-22.09.1962, at Lound 16.06.1968 and at Gunthorpe / Burton Joyce 12-14.09.1973.
The fifth county record of this North American wader was at Gunthorpe 08-09.06.1974. Since then another 20 have been found in Nottinghamshire with 4 birds in 2003, part of an exceptional influx into Britain.

Pectoral Sandpiper - birds recorded in Nottinghamshire 1975-2007

Years	1975-79	1980-84	1985-89	1990-94	1995-99	2000-04	2005-07	Total
Birds	4	4	1	4	1	4	2	20

Months	J	F	M	A	M	J	J	A	S	O	N	D
Birds	0	0	0	0	0	0	4	4	11	1	0	0

This species is essentially an early autumn vagrant with all records since 1975 falling between 17.07.1999 (Hoveringham) and 29.10.2007 (Lound). Most British records fall within this period.
Records are well spread with birds at 14 different sites in the county since 1974 including 9 different localities in the Trent Valley.

☐ **CURLEW SANDPIPER** *Calidris ferruginea*
The Curlew Sandpiper is primarily an uncommon autumn passage migrant through Nottinghamshire with records between July and October in all years from 1974 except 1987 when no birds were recorded in the county. However double figure counts are rare and there were only 8 site counts of 10 or more birds between 1974 and 2007. The maximum for the period is 32 at Stoke Bardolph floods on 09.09.1998 at a time when there was a large influx of birds into England.
This species is much rarer at other times. There were spring records (April - June) in 15 years between 1974 and 2007 but no site recorded more than 2 birds at one time. Within the same period there were 4 winter records - all of single birds in early November.
This is another species which was formerly more common in the heyday of Nottingham SF where 100 were recorded on 14.09.1946 and 28.09.1946.
Extreme dates 01.04.1996 (Hoveringham) - 09.11.1991 (Lound).

☐ **PURPLE SANDPIPER** *Calidris maritima*
The Purple Sandpiper is a coastal wintering species in Britain and a very rare late autumn visitor to Nottinghamshire with just 13 records in total including 9 birds seen in the Trent Valley. There were 7 between 1864 and

1972 - 1 killed on the River Trent near Wilford Ferry in summer 1864 and 1 dead at Larch Farm near Blidworth 25.09.1885 followed by singles at Nottingham SF 22.08.1944 and 09.08.1946, Netherfield 23.09.1949 and at Holme Pierrepont 18.09.1966 and 24.10.1972.

There have been 6 other birds since then:
1976   1 at Attenborough NR on 04.11.1976.
1983   1 at Lound on 07.11.1983.
1986   1 at Girton on 02-03.11.1986.
1996   1 at Lound on 20.09.1996.
2003   1 at Lound on 31.08.2003.
2004   1 at Netherfield Lagoons and Holme Pierrepont 21-24.11.2004.

In addition KA Naylor lists a record of one at Colwick CP 06.10.1980 although this record was never published in the Nottinghamshire Birdwatchers Annual Report.

☐ DUNLIN *Calidris alpina*

A fairly common passage migrant through Nottinghamshire with groups of up to 200 recorded at Nottingham SF before it was redeveloped. Numbers have been lower since then, perhaps a reflection of the limited habitat for migrating waders in recent years.

Dunlin - maximum site counts in Nottinghamshire 1975-2007

Years	1975-79	1980-84	1985-89	1990-94	1995-99	2000-04	2005-07
Maximum site counts	142	192	81	78	144	90	49
Counts of 100-200	6	2	0	0	1	0	0

Three-figure flocks have only been recorded 3 times since 1979 with 192 at Misson in March 1980, 100 at Misson in mid March 1981 and 144 at Holme Pierrepont 09.02.1996.

Two subspecies migrate through the county - the **Northern Dunlin** *C. a. alpina* and the **Southern Dunlin** *C. a. schinzii*.

☐ BROAD-BILLED SANDPIPER *Limicola falcinellus*

The Broad-billed Sandpiper is a rare vagrant to Britain which is rarely recorded inland. There is a single Nottinghamshire record of one at Nottingham SF on 23-26.08.1961.

☐ BUFF-BREASTED SANDPIPER *Tryngites subruficollis*

The Buff-breasted Sandpiper is a North American wader which has been recorded 5 times since 1975:
1975   1 at Netherfield 13-18.09.1975.
1978   1 at Gunthorpe 24-28.09.1978.
1986   1 at Bleasby 14-15.05.1986.
1992   1 at Netherfield 12-13.09.1992.
1995   1 at Lound 23-25.09.1995 moving to Girton 26.09.1995.

These are the only county records. Most British records involve autumn birds.
See NBAR 1986 p.44 and NBAR 1992 p.92.

☐ **RUFF** *Philomachus pugnax*
The Ruff is a moderately common passage migrant (March-May and July-October) and scarce winter visitor to Nottinghamshire.

Ruff - maximum site counts in Nottinghamshire 1996-2007

Years	96	97	98	99	00	01	02	03	04	05	06	07
November-February	4	1	9	3	23	5	3	7	3	0	4	0
March-June	6	42	4	8	2	6	17	8	2	2	1	3
July-October	9	11	35	7	18	9	14	10	7	2	9	4

42 at Lound 01.05.1997 is the highest published site count 1974-2007. However the highest count for Nottinghamshire is of 84 birds at Nottingham SF on 20.08.1944.

☐ **JACK SNIPE** *Lymnocryptes minimus*
The Jack Snipe is an uncommon passage migrant and winter visitor to Nottinghamshire. It is a secretive bird which spends a great deal of time concealed amongst reeds and other marsh vegetation. As a result it is probably under-recorded but on occasions reasonable numbers have been counted at favoured wintering sites as set out below:

Jack Snipe - annual site maxima in Nottinghamshire 1975-2004

Years		1975-1984	1985-1994	1995-2004
Annual site maxima	1-4 birds	1	3	1
	5-9 birds	8	7	5
	10-14 birds	1	0	3
	15+ birds	0	0	1

There were only 6 double-figure counts in Nottinghamshire between 1974 and 2007 - 10 at Bennerley Marsh 27.12.1995, 10 at Ollerton 04.02.2002, 10 at Langford Lowfields 04.12.2004, 12 at Huthwaite 02.01.1975, 12 at Ollerton 10.02.2003 and an exceptional 21 at Bennerley Marsh 19.03.1995. However, there is an earlier record of 27 together at Huthwaite 17.03.1971.
Records from 1995 show the seasonal distribution of Jack Snipe records in the county with even numbers of wintering birds between November and March:

Jack Snipe - birds recorded in Nottinghamshire 1995-2007

Months	J	F	M	A	M	J	J	A	S	O	N	D
Birds	191	207	223	49	2	0	0	0	15	129	214	174

Extreme dates 08.08.1976 (Colwick) - 01.06.1969 (Torworth).

☐ **COMMON SNIPE** *Gallinago gallinago*
Historically this species was a common winter visitor to Nottinghamshire and - until the 1930s - a fairly common breeding species in the county with reduced numbers breeding through to 1973. As a breeding bird the Common Snipe has done badly since then, a casualty of changes in land use which have reduced the amount of damp wetland sites available for breeding. A survey in 1982 found 61 pairs in the county but this total was reduced to c.8 pairs in 1992 and a key breeding site on pasture land in the Idle Valley was ploughed up in 1997. Since that date, drumming birds have been found at 1-6 sites per year.

The Common Snipe remains a reasonably widespread winter visitor but there have been few really large gatherings in recent years, reflecting the general decline:

Common Snipe - maximum site counts in Nottinghamshire 1975-2007

Year	1975-79	1980-84	1985-89	1990-94	1995-99	2000-04	2005-07
Maximum site count	600	350	153	210	212	240	70

600 birds recorded at Misson in March 1978 is the best count for 1974-2007.
See NBAR 1982 pp.31-33.

☐ **GREAT SNIPE** *Gallinago media*
There are 9 records of this vagrant snipe in Nottinghamshire. 4 were shot in the county before the First World War at Hickling 03.10.1882, near Southwell 1883, at Wollaton 1910 and at Besthorpe 03.10.1913.
4 more were recorded before 1974 at Nottingham SF 13.09.1947, at Huthwaite Flashes 16.11.1968, on the River Trent at Gunthorpe 03.01.1971 and at Hoveringham 30-31.03.1973.
One has been found since then:
1989   A juvenile at Girton 27.08-02.09.1989.
See NBAR 1989 pp.47-48.

☐ **LONG-BILLED DOWITCHER** *Limnodromus scolopaceus*
This North American vagrant has been recorded once in the county - a first-winter bird at Lound 21-28.10.1996. The majority of English records have been found between mid September and mid November.
See NBAR 1996 p.103.

◻ WOODCOCK *Scolopax rusticola*

A locally common resident species breeding in many older woodlands, particularly in the centre and north of the county. Birds are often unobtrusive except in the breeding season when display flights take place at dusk and dawn. However the species does move through the county in small numbers - particularly in March, November and December - and migrants are sometimes flushed in unusual locations such as Nottingham General Cemetery. Larger numbers are occasionally encountered at this time with - for example – a succession of double figure counts at Staunton (11 on 17.12.2002, 17 flushed during a pheasant shoot 13.11.2004, 11 on 30.12.2004 and 20 on 01.01.2006) and an exceptional count of c40 flushed at Farndon Willow Holt on 17.12.1978.

◻ BLACK-TAILED GODWIT *Limosa limosa*
This species is an uncommon passage migrant through Nottinghamshire which usually occurs in greater numbers than Bar-tailed Godwit. Greater numbers occur on autumn migration as set out below:

Black-tailed Godwit - double-figure site counts in Nottinghamshire 1975-2007

Months	J	F	M	A	M	J	J	A	S	O	N	D
Double-figure counts	0	0	1	8	0	0	10	8	3	0	2	0

Large flocks are occasionally recorded and there are 7 counts of 50 or more birds between 1974 and 2007 - 50 at Attenborough NR 28.08.1978, 78 in the Idle Valley 21.04.1983, 56 at Colwick CP 27.04.1998, 50 at Netherfield 17.11.2000, 64 at Colwick CP 27.04.2001, 63 at Lound 18.04.2005 and 86 west at Hoveringham 22.04.2006 (the highest count for 1974-2007).

2 subspecies migrate through the county - the **nominate form** *L. l. limosa* and the **Icelandic form** *L. l. islandica*.

□ BAR-TAILED GODWIT *Limosa lapponica*

The Bar-tailed Godwit is an uncommon passage migrant through Nottinghamshire.

There is distinct spring bias to records in the county as set out below:

Bar-tailed Godwit - double-figure site counts in Nottinghamshire 1975-2007

Months	J	F	M	A	M	J	J	A	S	O	N	D
Double-figure counts	0	1	0	6	4	0	2	1	3	0	0	0

This spring bias is confirmed by the pattern of occurrence over the last 13 years:

Bar-tailed Godwit - birds recorded in Nottinghamshire 1995-2007

Months	J	F	M	A	M	J	J	A	S	O	N	D
Birds	4	0	9	110	108	4	51	14	22	4	6	2

Large flocks are extremely unusual and there have only been 2 records of 40 or more birds since 1974 - 44 at Holme Pierrepont 20.04.1974 and 50 at Silverhill Tip 21.04.2001.

□ WHIMBREL *Numenius phaeopus*

The Whimbrel is an uncommon passage migrant, primarily between April and May with lesser numbers between July and September. However few birds ever land in the county with most preferring to fly straight through.

Whimbrel - maximum site counts in Nottinghamshire 1975-2007

Year	1975-79	1980-84	1985-89	1990-94	1995-99	2000-04	2005-07
March-June	5	4	7	67	9	28	5
Counts of 25+ birds	0	0	0	1	0	1	0
July-October	17	25	13	16	16	12	28
Counts of 25+ birds	0	0	0	0	0	0	1

Large groups were rare in the period 1974-2007 and there were only 3 counts of over 25 birds - an exceptional 67 birds at Lound on 07.05.1994, 28 at Eakring 10.05.2001 and 28 south at Farnsfield 01.08.2006.

Extreme dates 20.03.1976 (West Bridgford) - 23.10.1960 (Nottingham SF).

☐ **EURASIAN CURLEW** *Numenius arquata*
The Eurasian Curlew has enjoyed mixed fortunes in Nottinghamshire. It was a rare visitor in the Victorian era and only began to nest from 1944 onwards in small numbers (maximum c.14 pairs) as part of a general spread of this species into southern and eastern England in the first half of the 20th century. Outside the breeding season it was a frequent visitor to the old Nottingham SF prior to the redevelopment of the site with flocks of up to 400 birds from August through to October.
In the period 1974-2007 breeding numbers have remained small with a peak of c.17 territories in 1978. However fewer than 10 pairs have nested in recent years as lowland marshes have been drained and ploughed up. Passage numbers are also much reduced with few recent counts of 50 or more birds.

Eurasian Curlew - maximum site counts in Nottinghamshire 1975-2007

Year	1975-79	1980-84	1985-89	1990-94	1995-99	2000-04	2005-07
Maximum site count	200	47	37	56	54	56	55
Counts of 49+ birds	1	0	0	6	1	2	1

The highest count for 1974-2007 is of 200 birds west at Aslockton on 29.07.1977.

☐ **SPOTTED REDSHANK** *Tringa erythropus*
This species is primarily an uncommon annual passage migrant through Nottinghamshire with peak numbers occurring between July and October. Far smaller numbers are recorded between April and June and spring records are less than annual but they include the best recent site count for the county - 9 birds at Netherfield 30.04.2004. Winter records (November - February) are unusual but 1 or 2 birds were regularly reported in the Idle Valley in most winters between 1983 and 1992.

☐ **COMMON REDSHANK** *Tringa totanus*
The Common Redshank first bred in Nottinghamshire around 1884. In the first half of the 20th century the species was a common breeding species but it has declined considerably since the 1950s with changing land use patterns. A survey in 1958 revealed 100-113 pairs in the county but a decline set in after that date. Only 38-42 pairs were found in 1978 and 46 pairs were discovered in a full survey in 1982. By 2002 the species was confined as a breeding bird to just 10 sites.
Despite the decline in breeding numbers the Common Redshank remains a fairly common passage migrant and winter visitor with 120 at Misson/Newington in March-April 1978 and 93 at Girton 26.12.2000 - the best counts for 1974-2007.
See NBAR 1982 pp.31-33.

☐ GREENSHANK *Tringa nebularia*
The Greenshank is an uncommon passage migrant and rare winter visitor to Nottinghamshire with peak numbers occurring in August and September. Key sites for migrant birds include Hoveringham and Holme Pierrepont in the Trent Valley and Lound. Double-figure site counts are uncommon but underline the marked autumn peak in records of this species:

Greenshank - double-figure site counts in Nottinghamshire 1985-2007

Months	J	F	M	A	M	J	J	A	S	O	N	D
Double-figure counts	0	0	0	1	1	0	0	12	1	0	0	0

The largest count for 1974-2007 is of 19 birds at Lound on 19.08.2001.

☐ LESSER YELLOWLEGS *Tringa flavipes*
A bird shot at Misson in the winter of 1854-55 is the first British record of this North American wader.
There has only been one Nottinghamshire record since:
1995   A first-winter at Holme Pierrepont 25.11-15.12.1995 (which probably also visited Staffordshire and Shropshire).
See NBAR 1995 pp.114-115.

☐ SOLITARY SANDPIPER *Tringa solitaria*
1 at Nottingham SF 24.08-02.09.1962 is the only county record of this rare transatlantic vagrant. This was the seventh record for Britain.
See NBAR 2000 pp.141-142.

☐ GREEN SANDPIPER *Tringa ochropus*
A fairly common passage migrant and winter visitor with peak numbers occurring between July and September. Site counts for the period 2000-2007 underline the distinct autumn bias to records of this species:

Green Sandpiper - maximum site counts in Nottinghamshire 2000-2007

Years		00	01	02	03	04	05	06	07
Maximum	March-June	5	8	5	5	5	5	5	3
Site	July-October	12	11	13	8	20	13	23	10
Counts	November-February	6	7	10	5	5	4	4	4

Double figures counts are uncommon and 23 at Lound 18.08.2006 is the best count for 1974-2007.
The species was formerly recorded in large numbers in the heyday of the Nottingham SF with as many as 80-90 there on 09.08.1945.

☐ WOOD SANDPIPER *Tringa glareola*
The Wood Sandpiper is an uncommon passage migrant in very small numbers with peak numbers occurring in August and September. Records since 1990 underline the distinct autumn peak in records in Nottinghamshire:

Wood Sandpiper - birds recorded in Nottinghamshire 1990-2007

Months	J	F	M	A	M	J	J	A	S	O	N	D	Total
1990-94	0	0	0	2	17	4	5	c17	14	2	0	0	c61
1995-99	0	0	0	2	11	2	4	20	9	1	0	0	49
2000-04	0	0	0	0	7	2	10	38	6	0	0	0	63
2005-07	0	0	0	1	5	1	5	13	3	0	0	0	28
Total	0	0	0	5	40	9	24	c88	32	3	0	0	c201

The largest count for the period 1974-2007 is of up to 9 birds near Newington between July and September 1977.

The species formerly occurred in larger numbers with up to 60 birds passing through Nottingham SF annually from the 1940s until the 1960s. The lower numbers in recent years perhaps reflect the reduced habitat for passage waders in the county.

☐ **COMMON SANDPIPER** *Actitis hypoleucos*

The Common Sandpiper is a fairly common passage migrant and extremely rare breeding bird.

Breeding was suspected in 1973 and first proved in 1990.

Common Sandpiper - breeding records in Nottinghamshire 1974-2007

Decade	Year	Proven breeding	Possible breeding	Year	Proven breeding	Possible breeding
1980s	1980	0	2 pairs	1981	0	1 pair
1990s	1990	1 pair	0	1994	1 pair	2 pairs
	1997	1 pair	0	1999	1 pair	0
2000s	2000	2 pairs	0	2001	0	2 pairs
	2002	1 pair	0	2005	1 pair	0
	2007	1 pair	0	Total	9 pairs	7 pairs

Several breeding records have involved pairs nesting at abandoned pit tops in the west of the county (eg at Annesley Pit Top).

Spring and autumn passage numbers were fairly steady during the same period with annual site maxima of 6-18 birds between 1986 and 2007. However the largest count since 1974 is of 34 at South Muskham 29.07.1985. In contrast winter records (November - March) are less than annual.

☐ **SPOTTED SANDPIPER** *Actitis macularius*

There is only one record of this North American species - a single bird at Holme Pierrepont 17.12.1994-05.01.1995.

An earlier claim of 1 shot at Budby in 1848 is not now considered to be acceptable.

See NBAR 1994 pp.94-95.

☐ **TURNSTONE** *Arenaria interpres*

An uncommon passage migrant with peak numbers occurring in April to May and July to August. Counts of 1 to 5 birds are normal and there are only 4 double-figure counts for Nottinghamshire between 1974 and 2007 - 10 at Lound 20.08.1993, 12-14 at Netherfield 12.08.2000 (with another 5 birds there later the same day), 10 at Lound again 13.05.2006 and 14 at the same site on 17.05.2006. Prior to 1974 there was a large influx of over 100 birds in the autumn of 1944 including a county record of 22 on the River Trent 24.07.1944.

Winter records are scarce with only 8 records of 10 birds between November and January in the last 34 years.

□ WILSON'S PHALAROPE *Phalaropus tricolor*
There is one record for the county - a bird at Burton Meadows 16-28.07.1961. This was the seventh British record of this North American vagrant.

□ RED-NECKED PHALAROPE *Phalaropus lobatus*
The Red-necked Phalarope is an extremely rare visitor to Nottinghamshire with just 14 birds recorded in the county. The first bird was shot at Ramsdale 06.07.1843. This was followed by 5 records at the old Nottingham SF from 1947 - 1 24-31.08.1946, up to 4 birds 20-23.09.1947, 1 01.09.1961, 1 6-11.08.1963 and 1 30.08.1963.

Since 1974 only 5 further birds have been found in Nottinghamshire:
1983   1 at Reed Pond, Lambley in spring 1983.
       1 at Lound 25-27.09.1983.
1989   1 at South Muskham 30.09.1989.
1991   1 at Lound 07.06.1991.
1992   1 at Hoveringham 29-30.05.1992.

□ GREY PHALAROPE *Phalaropus fulicarius*
This species is a rare and infrequent late autumn visitor to Nottinghamshire with 14 records involving 15 birds c1870-1962. All dated records fell between September and November with the exception of 1 picked up dead at Kirton 16.07.1914.

There have been 9 or 10 birds since then:
1981   1 found dead at Beeston Rugby Field in early October 1981.
1983   1 at Lound 24.09.1983 with the same or another there 04-11.10.1983.
1987   1 at Burton Meadows 17.10.1987.
1996   1 at Colwick CP and Netherfield 03.11-12.11.1996.
1997   1 at Attenborough NR 21.10.1997.
       1 at Bennerley Marsh 30.11.1997.
2001   1 at Netherfield 01-03.10.2001.
2002   1 at Newington NT Reserve 29.10-04.11.2002.
2005   1 at Holme Pierrepont 06.11.2005.

◻ **POMARINE SKUA** *Stercorarius pomarinus*
There are two records of this species before 1970 - a bird killed at Clipstone in 1869 and another killed at Farnsfield in November 1875. A further bird was found in the Trent Valley at Holme Pierrepont on 20.06.1973.
3 more birds have been found in the Trent Valley since 1974:
1977 1 at Netherfield 12.11.1977.
1988 1 at Colwick CP 21.11.1988.
      1 at South Muskham 01.12.1988.

◻ **ARCTIC SKUA** *Stercorarius parasiticus*
The Arctic Skua is a rare passage visitor to Nottinghamshire. There is one 19th century record of this species - an immature killed between Farnsfield and Southwell one autumn sometime before 1866. A further 5 September birds were found in the Trent Valley between 1957 and 1973. Another was found at Attenborough NR 28.09.1974 and since then 37 more birds have been recorded:

Arctic Skua - birds recorded in Nottinghamshire 1975-2007

Years	1975-79	1980-84	1985-89	1990-94	1995-99	2000-04	2005-07	Total*
Birds	2	4	8	4	11	2	6	37

Months	J	F	M	A	M	J	J	A	S	O	N	D
Birds	0	0	1	0	4	5	0	14	7	3	2	1

* KA Naylor lists an additional record of 2 birds at Girton 12.05.1987 which was never published in the NBAR and is therefore not included in the totals.
Of the 38 records since 1974, 9 birds were found between 9th May and 29th June and 25 between 13th August and 16th October which suggests that a tiny overland passage may occur through the county. The fact that 27 of the 38 birds were found in the Trent Valley between Attenborough NR and Hoveringham also supports the theory that many records involve passage birds rather than those blown inland by storms at sea. All county records are of single birds apart from 3 at Nottingham SF 15.09.1957, 4 at Netherfield on the unusual date of 26.06.1997 and 5 at Holme Pierrepont 13.08.2006.
In addition to the above records there are 6 recent records of **unidentified Skuas** in Nottinghamshire in October 1978, October 1979, August 1980, May 1981, October 1987, December 1988 and November 2000. Several of these records probably involved Arctic Skua.

◻ **LONG-TAILED SKUA** *Stercorarius longicaudus*
There is one old and perhaps doubtful 19th century record of this species - 1 captured near Tuxford in early June 1881.
No doubts exist about the 3 more recent autumn records for Nottinghamshire:
1997 An immature bird at Colwick CP 13.09.1997.

1998   An approachable juvenile bird at Bentick Pit Top 10-12.08.1998.
2007   Another approachable juvenile in fields between Newton and Shelford Manor 17-30.09.2007.
See NBAR 1997 pp.139-140.

□ **GREAT SKUA** *Stercorarius skua*
There was one 19th century record of Great Skua in Nottinghamshire - a bird at Lamb Close on 22.08.1898. Singles followed at Netherfield and Holme Pierrepont 26-27.10.1968 and at Holme Pierrepont 10.10.1973 and there are 15 subsequent records between 1977 and 2006:

Great Skua - birds recorded in Nottinghamshire 1975-2007

Years	1975-79	1980-84	1985-89	1990-94	1995-99	2000-04	2005-07	Total
Birds	2	1	1	5	3	2	1	15

Months	J	F	M	A	M	J	J	A	S	O	N	D
Birds	2	1	0	2	1	0	0	1	3	5	0	0

The 15 recent records are widely scattered across the county rather than concentrated in the Trent Valley. Only 1 bird lingered for more than a day.

1977   1 found dead at Gringley Carr 05.04.1977 (ringed in Iceland).
1978   1 at King's Mill Reservoir 05.09.1978.
1983   1 at Gamston, Nottingham 19.02.1983.
1989   1 at Girton 23.04.1989.
1990   1 at Attenborough NR 28.01.1990.
1992   1 at Carlton 24.10.1992.
1993   1 at Lound 30.01-31.03.1993.
1994   1 at King's Mill Reservoir 23.05.1994.
       1 near Newark 05.10.1994.
1996   1 at King's Mill Reservoir 21.09.1996.
1998   1 at Colwick 02.10.1998.
1999   1 at Langar Airfield 09.08.1999.
2001   1 at Colwick 20.09.2001.
2002   1 at Huthwaite 31.10.2002.
2005   1 at Lound 26.10.2005.

□ **MEDITERRANEAN GULL** *Larus melanocephalus*
This species was added to the Nottinghamshire list in 1979 when an adult was found at South Muskham 13.01.1979. Singles were found in February 1983, July 1985 and January 1986 and there were 5 birds in 1987. Numbers then increased steadily with annual records from 1989 and between 17 and 30 birds were recorded in each year 1996-2007.

Mediterranean Gull - birds recorded in Nottinghamshire 1975-2007

Years	1975-79	1980-84	1985-89	1990-94	1995-99	2000-04	2005-07	Total
Birds	1	1	10	c38	c87	123	c82-87	c342-347

Many have been found in the gull roosts at Hoveringham and King's Mill Reservoir or amongst loafing gulls at Netherfield and Colwick CP.
No pairs have yet nested in Nottinghamshire but a bird summered at Lound for several years and hybridised with a Black-headed Gull in 1995.

☐ LITTLE GULL *Hydrocoleus minutus*
The Little Gull is an uncommon annual passage migrant through Nottinghamshire with a concentration of records in April and May. Numbers vary considerably from year to year with a low of 2 birds in 1976 and a high of 124 birds in 1984.

Little Gull - birds recorded in Nottinghamshire 1975-2007

Years	1975-79	1980-84	1985-89	1990-94	1995-99	2000-04	2005-07	Total
Birds	26	168	118	178	141	113	62	806

Months	J	F	M	A	M	J	J	A	S	O	N	D	Undated
Birds	4	7	4	298	300	16	20	57	31	20	23	19	7

There was one significant influx in the period 1974-2007 involving c97 birds through the county 01-14.05.1984. This influx included c92 birds through on 01.05.1984 with a minimum of 80 birds at Colwick CP and Holme Pierrepont.
A single pair laid eggs in the Black-headed Gull colony at Lound in June 1987 but the nest was later robbed. This was the fourth British breeding attempt.

☐ SABINE'S GULL *Larus sabini*
There are 4 fully acceptable county records of this autumn migrant - all of immature or juvenile birds. The first was at Nottingham SF 17-25.09.1950.
3 further birds have been found since 1974:
1982   1 at King's Mill Reservoir 01.10.1982.
1994   1 at Bennerley Marsh on 15.09.1994.
2004   1 at King's Mill Reservoir 25.09.2004.
September and October are the key months for records of this species in Britain.

☐ BLACK-HEADED GULL *Chroicocephalus ridibundus*
A very common non-breeding visitor and scarce breeding bird which first nested in Nottinghamshire in 1928. There are few regular breeding sites within the county but the species has nested on at least one occasion at 9 or more sites since 1994 - Hoveringham, South Muskham, Girton, Langford Lowfields, Idle Stop, Hallcroft, Bevercotes, Collingham Pits and Lound. However many of these sites are small and transitory. Lound currently holds the only large regular colony in the county with a peak of 726 pairs in 2001.

As a non-breeding bird the Black-headed Gull is found throughout the county with up to 25,000 birds forming large winter roosts on a number of waters (for example at Lound, King's Mill Reservoir and Girton).

Black-headed Gull - maximum site counts in Nottinghamshire 1975-2007

Years	1975-79	1980-84	1985-89	1990-94	1995-99	2000-04	2005-07
Site maxima	5000	5200	7000	8000	7000	10000	15000

As with other gull species, Hoveringham holds the largest roosting numbers in the county with up to 15,000 birds there on 30.12.2005 - the largest county total for a single site 1974-2007.

☐ RING-BILLED GULL *Larus delawarensis*

Two records from the 1990s are the only county records of this frequent North American visitor to Britain:
1990   A first-summer at Lound 24.03-25.04.1990 (with gaps).
1996   A first-summer at Gibsmere and Thurgarton 13.03- c01.04.1996.
See NBAR 1990 pp.52-54.

☐ COMMON GULL *Larus canus canus*

The Common Gull nested in the Trent Valley between 1967 and 1969 (and perhaps also in 1970) but there are no subsequent breeding records for Nottinghamshire.

Outside the breeding season it is a common winter visitor which regularly roosts on larger waters in the county. Between 1974 and 1989 the largest roost counts for the county were of 300 birds but numbers have increased steadily since then with a succession of four-figure counts at Hoveringham from 1998.

Common Gull - maximum site counts in Nottinghamshire 1975-2007

Years	1975-79	1980-84	1985-89	1990-94	1995-99	2000-04	2005-07
Site maxima	300	182	300	800	1000	4000	1000

A roost count of 4000 birds at Hoveringham on 13.01.2001 is the largest site count 1974-2007.

Birds showing characteristics of the race *L. c. heinei* were recorded 3 times at Colwick CP and once at Hoveringham between December 1997 and December 1998.

☐ LESSER BLACK-BACKED GULL *Larus fuscus graellsii*

A common passage migrant and wintering bird in Nottinghamshire where numbers have increased considerably from the 1970s.

Lesser Black-backed Gull - maximum site counts in Nottinghamshire 1975-2007

Years	1975-79	1980-84	1985-89	1990-94	1995-99	2000-04	2005-07
Site maxima	740	800	1100	1675	800	3000	2400

A massive 3000 at Hoveringham on 16.10.2002 is the highest count for 1974-2007.

Passage and wintering birds have been described with the characteristics of the form *L. f. intermedius*.

The Lesser Black-backed Gull is an extremely rare breeding bird in the county with successful breeding in 1963-64 and 2007 when 2 pairs bred at Lound. Pairs also attempted to breed in 1945, 1960-61, 1990 and 2004.

☐ **HERRING GULL** *Larus argentatus argenteus*

A common passage migrant and wintering bird in Nottinghamshire, where numbers have increased considerably since the 1970s.

Herring Gull - maximum site counts in Nottinghamshire 1975-2007

Years	1975-79	1980-84	1985-89	1990-94	1995-99	2000-04	2005-07
Site maxima	550	400	800	900+	1000	3000	4500

A massive 4500 in the large mixed gull roost at Hoveringham on 15.01.2005 is the highest count 1974-2007. A small number of wintering birds belong to the subspecies *L. a. argentatus* and a bird at Wollaton Park on 25.11.1995 was described as showing characteristics of the form *L. a. omissus*.

Very small numbers of Herring Gulls are present in the summer but no breeding attempts have so far taken place.

☐ **WESTERN YELLOW-LEGGED GULL** *Larus michahellis*

This species - for a long time considered a race of Herring Gull - was first recorded in Nottinghamshire at Attenborough NR on 26.07.1981. Small numbers have been found in every year since then. Now that the identification criteria are more clearly established, the species has been shown to be a scarce but regular passage migrant and rare winter visitor with records at 8-21 sites per year since 1992. There is a distinct build up in the number of birds present between July and October and 10 at Cotham Tip 19.08.2003 and 10 at Kilvington New Lake on 10.08.2007 are the best counts for Nottinghamshire.

☐ **CASPIAN GULL** *Larus cachinnans*

Birds 'showing characteristics' of Caspian Gull present a new challenge for seasoned gull watchers and the true status and distribution of this species is only just beginning to emerge as identification criteria are sorted out.

Birds have been recorded annually since the first was identified at Hoveringham 23-24.12.1997.

Caspian Gull - birds recorded in Nottinghamshire 1997-2007

Years	97	98	99	00	01	02	03	04	05	06	07
Birds	1	6	2	7	8	6	5	3	c9	c13	3

Months	J	F	M	A	M	J	J	A	S	O	N	D
Birds	17	9	2	0	0	1	0	3	0	2	9	20

34 of the c63 birds found to date have been identified at Hoveringham.
A particularly popular and instructive bird was recorded at Clumber, Welbeck and Carburton for 3 successive winters from 2003-04 to 2005-06.

☐ **ICELAND GULL** *Larus glaucoides glaucoides*
A scarce, winter visitor to Nottinghamshire. There were just 7 records of 10 birds between 1946 and 1973 (all found between 23rd November and 2nd May).
Since then many have been found by dedicated gull watchers at tips (for example at Dorket Head and Huthwaite) and at winter roost sites. Following 1 at Hoveringham 07.12.1974 a further 98 birds were recorded up to 2007:

Iceland Gull - birds recorded in Nottinghamshire 1975-2007

Years	1975-79	1980-84	1985-89	1990-94	1995-99	2000-04	2005-07	Total
Birds	1	2	4	12	37	28	14	98

Months	J	F	M	A	M	J	J	A	S	O	N	D
Birds	40	21	10	3	0	0	0	0	0	0	1	23

As with Glaucous Gull the roost at Hoveringham has been particularly productive with 43 birds recorded between 1995 and 2007.
Extreme dates 23.11.1973 (Mansfield) - 02.05.1970 (Attenborough NR) & 02.05.1997 (Lound).
In addition to the above records, the subspecies (or hybrid form) **Kumlien's Gull** *L.g.kumlieni* has been recorded once in the county - at Hoveringham 26.02.2006.
See NBAR 2006 pp.110-111.

☐ **GLAUCOUS GULL** *Larus hyperboreus*
A scarce winter visitor to Nottinghamshire with one 19th century record of 2 birds shot near Beeston Weir 22.12.1872 and 9 more birds between 1946 and 1973 (all found between 14th November and 30th April).
Since 1974 records have increased considerably as observers have combed large gatherings of gulls at refuse tips (for example at Huthwaite and Dorket Head) and roosts at lakes and reservoirs in search of rarer gulls.

Glaucous Gull - birds recorded in Nottinghamshire 1975-2007

Years	1975-79	1980-84	1985-89	1990-94	1995-99	2000-04	2005-07	Total
Birds	8	11	9	15	22	33	20	118

Months	J	F	M	A	M	J	J	A	S	O	N	D
Birds	41	28	18	3	1	0	0	0	0	2	6	19

The large and well watched roost at Hoveringham has consistently been the best site for this species with at least 40 birds recorded between 1995 and 2007 including 9 different birds there in 2001.
Extreme dates 15.10.1991 (Rainworth) - 15.05.1993 (Lound).

□ **GREAT BLACK-BACKED GULL** *Larus marinus*
In the early 1970s this species was a fairly common visitor to Nottinghamshire which had increased in the county since the 1960s. This trend has continued in the period 1974-2007 and the species is now a common winter visitor and uncommon summer visitor.

Great Black-backed Gull - maximum counts in Nottinghamshire 1975-2007

Years	1975-79	1980-84	1985-89	1990-94	1995-99	2000-04	2005-07
Site maxima	800	600	400	831	505	2000	4000

Hoveringham has been a key site for large roosts of this species in recent years with up to 2000 birds in December 2003 and January 2004 and an unprecedented 4000 on 18.12.2006. Lound also had over 1000 roosting birds in January 2004.

□ **KITTIWAKE** *Rissa tridactyla*
An uncommon and erratic annual visitor to Nottinghamshire, particularly in early spring.

Kittiwake - birds recorded in Nottinghamshire 1975-2007

Years	1975-79	1980-84	1985-89	1990-94	1995-99	2000-04	2005-07	Total
Birds	37	36	280	111	113	45	14	636

Months	J	F	M	A	M	J	J	A	S	O	N	D
Birds	34	23	377	64	26	10	6	18	9	19	34	16

The number of birds each year has varied from 1 to 196 birds as set out below:

Kittiwake - number of birds recorded each year in Nottinghamshire 1975-2007

Birds per year	1-10	11-20	21-30	31-40	41-50	51-99	100+
Number of years	16	9	4	1	2	0	1

1988 was the exceptional year for this species with 196 birds (44 in 1996 is the next best) and included 180 through Rainworth 07-14.03.1988 with 83 together there on 12.03.1988.These birds were part of an exceptional passage movement through the English Midlands (together with over 1000 birds in the West Midlands region) probably caused by a stationary weather front which interrupted the normal passage of this species. No doubt Whitaker would have delighted in the Rainworth record.

☐ GULL-BILLED TERN *Gelochelidon nilotica*
There is one old county record - a group of 5 birds at Netherfield on 06.09.1945.
One has occurred since:
2006   1 at Lound 19.05.2006.

☐ CASPIAN TERN *Hydroprogen caspia*
The Caspian Tern is a rare but regular summer vagrant to Britain and 11 have reached Nottinghamshire. The only 19th century record is of a bird shot on the River Trent at Colwick on 03.08.1894. There were 3 further records before 1974 - at Holme Pierrepont 11.06.1966, on the River Trent at Beeston Weir 15.08.1966 and at Attenborough NR 08.06.1970.
Since then there have been 7 further records:
1976   1 at Besthorpe 02.07.1976.
1988   1 at South Muskham and Winthorpe 04-12.05.1988.
        Another at South Muskham 12-13 and 28.07.1988.
1989   1 at Colwick CP 28.05.1989 and on the River Trent at Cottam 06.06.1989.
1993   1 at Lound 12.05.1993.
1998   1 at Lound and Hallcroft 03.07.1998.
1999   1 at Colwick CP 11.06.1999.
A large number of British records are concentrated in the period between June and August.
See NBAR 1988 pp.38-40 and NBAR 1993 pp.91-92.

☐ SANDWICH TERN *Sterna sandvicensis*
An uncommon but annual passage migrant with a maximum of 32 birds recorded in any one year (1990) since 1974. In that year, 17 were at Attenborough NR on 22.09.1990 - the highest count for the period 1974-2007. The county record is of an exceptional 53 west along the River Trent near Nottingham SF 18.09.1944.

Sandwich Tern - birds recorded in Nottinghamshire 1975-2007

Years	1975-79	1980-84	1985-89	1990-94	1995-99	2000-04	2005-07	Total
Birds	55	33	53	106	64	22	15	348

Months	J	F	M	A	M	J	J	A	S	O	N	D
Birds	0	0	4	81	67	21	32	38	102	1	0	2

Extreme dates 20.03.1958 (Toton) - 28.10.1976 (Netherfield) with an exceptional record of 2 birds at Netherfield and Colwick 06.12.1995.

▢ **ROSEATE TERN** *Sterna dougallii*
This rare breeding tern has reached Nottinghamshire 8 times but no bird has stayed longer than one day. Four were recorded in the Trent Valley before 1974 - at Nottingham SF 21.05.1945, Holme Pierrepont 11.05.1966, Netherfield 26.07.1970 and Netherfield again 13.08.1970.
Five or six more birds were recorded from 1974 including the first record away from the Trent Valley at Lound:
1986   1 at Netherfield 27.06.1986.
1990   1 at Hoveringham 07.05.1990.
2003   2 or possibly 3 at Hoveringham 10.09.2003.
2004   1 at Lound 06.06.2004.
A controversial bird, presumed to be a hybrid with Common Tern, returned to Colwick CP, Holme Pierrepont and Netherfield between 1992 and 1995.
For a discussion of the 1992-1995 bird see NBAR 1992 pp.88-89 and Birding World vol. 6/3 (March 1993) p.125 and vol 8/8 (August 1995) pp.310-311.

▢ **COMMON TERN** *Sterna hirundo*
The Common Tern has been an uncommon summer breeder in Nottinghamshire since 1945 but breeding numbers have increased steadily from 5-13 breeding pairs in the 1960s and early 1970s.

Common Tern - maximum breeding pairs in Nottinghamshire 1975-2006

Years	1975-79	1980-84	1985-89	1990-94	1995-99	2000-04	2005-06
Annual pairs	11-60	30-42	43-52	55-94	50-87	48-93	56+

The species has benefited from the extraction of gravel in the Trent Valley and elsewhere which has created many suitable breeding sites. Pairs bred at 17 different sites between 1994 and 2007 with up to 24 pairs breeding regularly at Colwick CP and large numbers at Attenborough NR peaking at 50-60 pairs in 2003.
Large numbers also pass through on migration with peak counts of 120 birds at Holme Pierrepont 02.05.1996 and again on 15.05.1998.
Extreme dates 24.03.2003 (Eakring) - 14.11.1960 (locality unrecorded).

▢ **ARCTIC TERN** *Sterna paradisaea*
The Arctic Tern is an uncommon passage migrant most years but is sometimes present in large numbers during spring migration. Annual numbers have ranged from a mere 3 birds in 1975 to an unprecedented 882-1388 through the county 01-03.05.1998. This movement included 200 through Colwick CP 02.05.1998 and 303 through there in 8 groups the next day. Similar numbers went through other parts of the midlands at the same

time. There is one other large spring count between 1974 and 2007 - 200 birds at Holme Pierrepont on 07.05.1991.

The Arctic Tern is much scarcer during autumn migration with only 179 birds in the last 18 years:

Arctic Tern - Late June to November records in Nottinghamshire 1990-2007

Years	1990	1991	1992	1993	1994	1995	1996	1997	1998
Birds	4	3	9	29	6	6	4	46	4
Years	1999	2000	2001	2002	2003	2004	2005	2006	2007
Birds	1	10	1	3	6	8	11	6	22

Extreme dates 11.04.1990 (Holme Pierrepont)–13.11.2005 (Hoveringham).

□ LITTLE TERN *Sternula albifrons*

A very scarce passage migrant through Nottinghamshire. Most records involve birds passing through between April and June. During the period 1974-2007 the most in any one year was 15 birds (1997) and there were none at all in 1981 and 1986.

Little Tern - birds recorded in Nottinghamshire 1975-2007

Years	1975-79	1980-84	1985-89	1990-94	1995-99	2000-04	2005-07	Total
Birds	30	19	24	23	31	27	5	159

Months	J	F	M	A	M	J	J	A	S	O	N	D	Undated
Birds	0	0	0	20	79	21	7	11	11	0	0	0	10

10 birds were in the county on 03.05.1997 including the largest group in the period - 7 birds at Attenborough NR. 6-7 were also at Attenborough NR on 05.09.1970.

There may be a small but regular passage of this species through the Trent Valley. For example Attenborough NR had a total of 34-36 birds between 1960 and 1973 and 37 more birds have been found there since 1974. Many other records of this species have come from nearby sites in the lower Trent Valley such as Netherfield, Holme Pierrepont and Hoveringham.

Extreme dates 20.04.1993 (Hoveringham) & 20.04.1997 (Lound) - 05.10.1944 (River Trent at Bulcote). In addition Whitaker makes a passing reference to two 19th century December records for the county.

□ WHISKERED TERN *Childonias hybrida*

There are 2 county records:
2002   A brief bird at King's Mill Reservoir on 02.05.2002.
2003   A well twitched bird at Hallcroft GPs and Lound (which had earlier been seen in East Yorkshire) 15-24.06.2003.

See NBAR 2003 pp.109-110.

□ BLACK TERN *Childonias niger*

An uncommon passage migrant through the county particularly in late spring when significant groups sometimes move through. The strongest passage movement involved 185 birds on 02.05.1990 and 190 birds on 03.05.1990 including c77 at South Muskham on the latter date. These birds formed part of a large weather related movement through the English Midlands. There was also a large autumn influx of over 300 birds 11-12.09.1992 which brought over 100 birds to both Attenborough NR and Holme Pierrepont.

The species has long since ceased to breed regularly in Britain but exceptionally a pair attempted to breed at Attenborough NR in 1978. The eggs of the pair were stolen and none have attempted to breed in Britain since then. The ♂ also summered alone in 1976-77 and 1979-80.

Extreme dates 12.04.1963 (in the Trent Valley) - 05.11.1984 (Stoke Bardolph).

☐ **WHITE-WINGED BLACK TERN** *Childonias leucopterus*
There are 10 county records. The first 3 were at Nottingham SF on 02.09.1945, 28.08.1946 and 14.09.1952.
There have been a further 7 birds since 1992:
1992   1 at Lound 16.05.1992.
        Another at Lound 11.09.1992.
1994   1 at Lound 05.08.1994.
        1 at Colwick CP, Netherfield and Holme Pierrepont 24.09-09.10.1994.
1996   1 at South Muskham 01-03.09.1996.
2004   1 at Girton 20.07.2004.
2006   1 at Thrumpton 13.09.2006.
See NBAR 1992 pp.86-88 and pp.90-91.

☐ **COMMON GUILLEMOT** *Uria aalge*
There are three curious county records of this species. The first involved several shot on the lake at Thoresby in late December 1855.
Two more have been recorded since 1974:
1995   1 flying over Rampton on 17.09.1995.
2007   An immature on the River Trent at Cottam 24.08.2007.

☐ **RAZORBILL** *Alca torda*
There are 5 records of this species for Nottinghamshire. The first was killed on the River Trent near Newark in January 1847 and another was killed on a canal near Eastwood in 1870.
Three more have been recorded since 1974:
1984   1 at Holme Pierrepont rowing course 11-14.02.1984.
1992   1 at Lound 07-18.06.1992.
1995   1 flying west over the A1 near Cromwell 27.09.1995.

☐ **LITTLE AUK** *Alle alle*
The Little Auk is a rare storm driven vagrant to Nottinghamshire. 36 birds were found in the county from 1841 to 1969 with all dated records occurring between 4th November and 20th February. These included 17 birds in February 1912.
Since 1969 only 9 further birds have been found:
1977   1 found dead at Farnsfield 24.11.1977.
1983   1 at Colwick CP 12.02.1983.
1984   1 at Farnsfield 09.11.1984 (which was found dead 3 days later).
1988   1 at Daneshill Reservoir 31.10.1988.
1989   1 at Colwick CP 15.11.1989.
1990   1 at Tuxford 08.12.1990.
1991   1 picked up at Hills and Holes late autumn 1991.
1995   1 flying over the A57 near Worksop 29.10.1995.
          1 picked up near Bingham 04.11.1995.

☐ **PUFFIN** *Fratercula arctica*
There are 9 records of this seabird in Nottinghamshire with 7 records involving dead or exhausted birds. 3 were recorded before 1918 with single birds picked up alive near Mansfield Woodhouse on 12.11.1884, killed near Bothamsall in autumn 1889 and picked up exhausted at Ramsdale 29.11.1917. Thereafter singles were picked up alive near Stoke Bardolph 07.11.1944 and near Oxton 09.10.1953. The latter bird is now on display at Wollaton Hall Natural History Museum.
At least four have been recorded since 1974:
1975   1 vigorous bird flew upstream between Burton Meadows and Burton Joyce 04.12.1975.
1983   1 found long dead at Idle Stop Pumping Station 11.05.1983.
1991   1 found dead at Ratcliffe-on-Soar Power Station 02.07.1991.
1992   1 was picked up exhausted at Mansfield Woodhouse 26.05.1992.
No doubt several of these birds were victims of autumn and winter storms.

☐ **PALLAS'S SANDGROUSE** *Syrrhaptes paradoxus*
This species erupted westwards to Britain several times in the late 19th and early 20th centuries and reached Nottinghamshire in 3 years 1863-1889:
1863   7 birds near Farnsfield and Mansfield between 01.06.1863 and 20.08.1863.
1888   c.76 birds found at Clifton, Costock, Besthorpe, Clipstone, Vicar's Pond, near Blidworth, Elmsley Lodge near Rainworth (60 April-September) and Cropwell Bishop between 22.05.1888 and 10.11.1888.
1889   1 bird at Rainworth in late May.
1888-89 saw the largest national invasion of this species and Whitaker's account of the invasion in Nottinghamshire contains a claim that at least one pair bred at Elmsley Lodge in June 1888.

☐ **ROCK DOVE** *Columba livia*
There are now a large number of feral domestic pigeons in Nottinghamshire, particularly in urban centres such as Nottingham, Mansfield, Newark, Retford and Sutton-in-Ashfield. Few comprehensive counts have appeared in the county annual reports but there are reports of flocks of 300-1500 in each year 1999-2007 with a maximum count of 1500 in fields at Toton on 16.11.1999.

☐ **STOCK DOVE** *Columba oenas*
A reasonably common breeding species in Nottinghamshire with flocks of up to 700 recorded in the county stronghold in the Idle Valley in recent years. Even larger numbers were recorded at 2 other sites in the 1990s with 800-1000 west through Colwick CP on 02.11.1990 and a regular winter roost at Besthorpe which held 500-1500 birds in 1996.

☐ **WOOD PIGEON** *Columba palumbus*
The Wood Pigeon is a very common resident species, passage migrant and winter visitor. Winter flocks of between 1000 and 3000 were recorded reasonably frequently between 1974 and 1991. However from 1992 to 2000 much larger flocks were recorded in the county, particularly in the Idle Valley where an enormous total of 21,400 birds were counted on 28.11.1993 with 10,000 birds in a single flock. 10,000 birds were also recorded roosting in the Idle Valley 22.01.1994. Since 2000 flocks seem to have returned to 1970s levels.

Wood Pigeon - maximum counts in Nottinghamshire 1975-2007

Years	1975-79	1980-84	1985-89	1990-94	1995-99	2000-04	2005-07
Site maxima	2000	2000	3000	21400	7020	5000	5000

☐ **COLLARED DOVE** *Streptopelia decaocto*
The Collared Dove was first recorded in the county in 1958 and first bred in 1959 at Osberton. Numbers have increased rapidly since then and this species is now a common resident throughout Nottinghamshire. Between September and January large groups gather together to feed and to roost in the county.

Collared Dove - maximum flock sizes in Nottinghamshire 1975-2007

Years	1975-79	1980-84	1985-89	1990-94	1995-99	2000-04	2005-07
Maximum site counts	300	400	327	220+	365	193	156
Flocks of 100-199	9	8	6	4	11	10	3
Flocks of 200-400	6	15	2	2	2	0	0

The largest count for Nottinghamshire is of 400 roosting at Attenborough NR 14.11.1981. However no flock larger than 193 birds has been recorded

1998-2007 and this may indicate a slight recent reduction in the number of birds in the county.

☐ **TURTLE DOVE** *Streptopelia turtur*
In the early 1970s the Turtle Dove was described as a fairly common passage migrant and summer visitor to Nottinghamshire which was first proved to breed in the county at Ramsdale in 1868. However it has declined considerably as a breeding bird in the county since the late 1980s and it is now an uncommon bird in Nottinghamshire. A survey of the species in 1998 found just 121 potential breeding birds.

Large numbers (up to 90 birds) were recorded a number of times between 1974 and 1997 and there were c300 at Misson 02.06.1982. However it is a reflection of the decline of this species that few large gatherings have been recorded since the early 1980s.

Turtle Dove - maximum flock sizes in Nottinghamshire 1975-2007

Years	1975-79	1980-84	1985-89	1990-94	1995-99	2000-04	2005-07
Flocks of 20-49	1	2	5	2	1	1	0
Flocks of 50-99	1	2	0	0	2	0	0
Flocks of 100+	0	1	0	0	0	0	0

Extreme dates 01.04.1981 (Attenborough NR) - 01.11.2004 (Caunton).
See *NBAR 1998 pp.155-161.*

☐ **ROSE-RINGED PARAKEET** *Psittacula krameri*
The Rose-ringed Parakeet has become established as a feral breeding species in southern England since the early 1970s.

The first Nottinghamshire record is of an escaped bird at East Bridgford on 16.01.1968. Since 1980 there have been an increasing number of records for Nottinghamshire but no pattern of occurrence has yet emerged and it is unclear how many of the birds originate from local cages and how many are wanderers from established populations. The pattern is also confused by misidentification of other escaped parakeet species.

Rose-ringed Parakeet - birds recorded in Nottinghamshire 1975-2007

Years	1975-79	1980-84	1985-89	1990-94	1995-99	2000-04	2005-07	Total
Birds	0	8	2	7	6	16	2	41

Months	J	F	M	A	M	J	J	A	S	O	N	D
Birds	3	4	4	4	4	1	1	8	4	4	2	2

Most records have involved short staying birds with only 10 birds since 1974 staying more than 1 day and only 2 long staying birds - singles at Stoke Bardolph and Shelford 14.08.1981 to mid November 1981 and at Ruddington 08.12.1994 - 28.02.1995. Perhaps the most intriguing record in this respect is of 7 flying through Bestwood CP 28.08.2003.
See *NBAR 2005 pp. 104-106.*

☐ **COMMON CUCKOO** *Cuculus canorus*
A fairly common but declining summer visitor throughout the county. Many birds are found at reclaimed gravel pits where eggs are laid in nests of Reed and Sedge Warblers. Birds normally arrive in April and most have drifted back south by late August - early September.
Occasionally small groups are seen together on migration and there was an exceptional gathering of 30 at Hoveringham 10-11.05.1981.
Extreme dates 25.03.1962 (Holme Pierrepont) - 18.11.1956 (Westview Farm, Besthorpe).

☐ **BARN OWL** *Tyto alba alba*
An uncommon resident breeding species in the county which has undergone a considerable decline in recent decades as a result of agricultural changes, the redevelopment of old barns and road deaths. To some extent this decline has recently been reversed by extensive nest box schemes run by the North Notts Ringing Group, the Rushcliffe Barn Owl Project, the South Notts Ringing Group and the Hawk and Owl Trust (now the Wildlife Conservation Partnership). As a result of this work there were estimated to be 100 pairs in the county in 2004 and over 500 young fledged in Nottinghamshire in 2007.

Barn Owl - annual number of sites in Nottinghamshire 1975-2004

Years	1975-79	1980-84	1985-89	1990-94	1995-99	2000-04
Sites reporting Barn Owls	14-24	15-25	8-22	17-24	19-24	15-40

There are 2 records of what were either wild or released **dark-breasted continental birds** *T. a. guttata* at Attenborough NR 23.05.1974 and at Gringley Carr 27-29.12.1996.
See NBAR 2005 pp.110-115 and NBAR 2006 pp. 126-128.

☐ **EURASIAN SCOPS OWL** *Otus scops*
One found at Bestwood CP on 28.06.1973 died shortly after it was found. It is the only county record and the only 20th century record of this southern European owl for the whole of the English Midlands.
See NBAR 1999 p.133-134.

☐ **LITTLE OWL** *Athene noctua*
This introduced bird was first recorded in Nottinghamshire when one was caught near Newark in 1896 as birds spread from nearby release sites in Northamptonshire and Yorkshire. It was first proved to breed in the county at Gonalston in 1913 and thereafter became well established in the county.
It remains fairly common today and favourable farmland areas can support large number of birds as shown by a local survey which revealed 36 pairs within a 20km radius of Walkeringham in 1977. Although widespread, the Idle Valley is the county stronghold for this species.

☐ **TAWNY OWL** *Strix aluco*

A widespread bird throughout the county which is more often heard calling than it is seen. It occurs in a wide variety of urban and rural woodland habitats and breeding densities can be high in mature woodland. For example, 15 pairs have been estimated to breed at Sherwood Forest CP in some years. However there has been some limited evidence to suggest that there has been a recent decline in numbers within Nottinghamshire.

☐ **LONG-EARED OWL** *Asio otus*

An uncommon and local breeding species and winter visitor to the county. It has been estimated that there were 30-50 breeding pairs in Nottinghamshire in the early 1970s and 20-30 pairs were found to be breeding regularly in a 210 square kilometre survey area of the River Idle and River Trent in the early 1990s. There are more recent estimates of at least 15 potential breeding pairs at 11 localities in 1998 and of 20 breeding pairs in 2005. However, the species is inconspicuous and difficult to properly assess so breeding numbers may still be significantly higher.

The Long-eared Owl is also prone to disturbance at winter roosts and little detailed information is therefore published about winter numbers. However there were large reported roosts of 15 birds on 26.10.1984 and 22 birds in 2001 at unnamed sites.

☐ **SHORT-EARED OWL** *Asio flammeus*

The Short-eared Owl is primarily an uncommon winter visitor and passage migrant to Nottinghamshire. Numbers vary considerably from year to year

with a low of 6 birds recorded in 1977 and a high of up to 76 birds in 1979. However, overall wintering numbers seem to have declined since the 1990s.
Occasional birds summer and the Short-eared Owl has bred 4 times in the county (at Rainworth in 1908 and at undisclosed localities in 1978, 1981 and 1999). Another pair lost eggs in the Idle Valley in 1973.

☐ **EUROPEAN NIGHTJAR** *Caprimulgus europaeus*
Nottinghamshire - along with Staffordshire - has the strongest population of this uncommon and local summer visitor in the English Midlands. In the early 1970s the county breeding population was estimated to be around 60 pairs centred on the sandy heaths and early growth forestry plantations in the west and central north-west of the county (for example at Budby, Clipstone Forest, Clumber Park and Blidworth Woods). There has been some shrinkage of range but not of numbers since then with 67 churring ♂s found during a full survey of the species in 1981. In more recent years birds bred at 15 different sites in the county between 1997 and 1999 and survey work revealed 66 churring birds in 2004 and 61 calling at 11 sites in 2005.
Extreme dates 21.04.1847 (locality unrecorded) - 10.10.1996 (roosting in a Bingham garden). There is one other October record for the county of a bird roosting in a Keyworth garden 01.10.1995.
See NBAR 2005 pp. 102-103.

☐ **EGYPTIAN NIGHTJAR** *Caprimulgus aegyptius*
There is one exceptional record of this species - the first for Britain - a bird shot at Thieves Wood near Mansfield on 23.06.1883. The only other British record is of a bird at Portland Bill, Dorset on 10.06.1988.
See NBAR 2001 pp.122-123.

☐ **COMMON NIGHTHAWK** *Chordeiles minor*
An immature bird hawking around Bulcote on 18 & 21.10.1971 is the sole county record of this rare North American vagrant. This was the fifth for Britain.

☐ **ALPINE SWIFT** *Apus melba*
One at Bennerley Marsh for 10 minutes on 01.09.1998 is, perhaps surprisingly, the only county record of this European vagrant.
See NBAR 1998 p.163.

☐ **COMMON SWIFT** *Apus apus*
A common summer visitor which traditionally arrives a fraction later than most migrants (at the end of April to early May) and departs early (between mid August and early September). However the screaming flocks of this species are one of the key sounds of high summer. Many of the really large spring and early summer flocks of this species build up in the Trent Valley

and Holme Pierrepont would appear to be a significant staging site on migration with 5000 birds on 21.05.1994, 14.06.1998 and 13.06.2000. Extreme dates 03.04.1992 (South Muskham) - 12.11.1952 (locality unrecorded).

☐ LITTLE SWIFT *Apus affinis*
The first and only county record of this rare vagrant to Britain was a widely appreciated bird at Netherfield 26-29.05.2001. There are 21 British records. *See NBAR 2001 pp.111-112.*

☐ COMMON KINGFISHER *Alcedo atthis*
A fairly common and reasonably widespread resident species in Nottinghamshire found on a range of lakes, rivers and streams. As an example, birds were recorded at 87 different localities during 2002. Many of the gravel pit complexes in the Trent Valley hold several pairs (eg. Colwick CP, Holme Pierrepont, Hoveringham and Attenborough NR all held at least 3 pairs in 1999). However Common Kingfishers suffer high mortality in severe winter weather conditions such as in 1962-63 and the population can take several years to recover.

☐ EUROPEAN BEE-EATER *Merops apiaster*
The only certain record for Nottinghamshire of this southern European vagrant was a bird at Huthwaite 27-28.08.1970.

☐ EUROPEAN ROLLER *Coracias garrulus*
1 rather controversial record - a bird at Babworth 28.06-22.07.1966 - was accepted by the BBRC but may have been a caged bird which escaped from Twycross, Leicestershire in June of the same year.

☐ HOOPOE *Upupa epops*
The Hoopoe has always been a rare and exciting visitor to the county with 14 to 16 birds recorded in Nottinghamshire from 1863 to 1972.
2 further birds were found in 1974 (at Tuxford in May 1974 and at Thorpe, Wysall 24.10.1974) and approximately 25 more were recorded to 2007:

Hoopoe - birds recorded in Nottinghamshire 1975-2007

Years	1975-79	1980-84	1985-89	1990-94	1995-99	2000-04	2005-07	Total
Birds	4	9	2	6	0	1	3	25

Months	J	F	M	A	M	J	J	A	S	O	N	D
Birds	1	0	0	7	7	6	1	0	0	1	2	0

Surprisingly only 4 have occurred since 1994 - a bird at Longdale Lane and Sansom Wood 02-09.06.2001 and singles at East Bridgford c07-09.04.2005, Kirkby in Ashfield 05.06.2005 and at Sutton Bonnington and Ratcliffe-on-Soar 08-21.04.2006.

As might be expected 21 records from 1974 involve overshooting spring migrants and April and May are the peak months for records throughout Britain. Four other records involve very late autumn migrants.
Extreme dates c07.04.2005 (East Bridgford) - 25.11.1990 (Sutton-on-Trent). In addition there is one exceptional winter record of a bird which was eventually found dead at Forest Town, Mansfield 15.01-19.02.1991.
See NBAR 2006 pp. 103-109.

□ **WRYNECK** *Jynx torquilla*
The Wryneck probably bred in 19th century Nottinghamshire when it was familiar enough to be known locally as the 'Cuckoo's Maiden' or the 'Cuckoo's Mate' and was suspected of nesting in the east of the county in 1944 and 1954-55. However, this species is now extinct as a regular breeding bird in England and - apart from a singing bird at Clipstone Forest 13-16.05.1989 and another calling at Lound 08.07.2005 - it has become a very scarce migrant visitor to Nottinghamshire.
After a single bird in a garden at South Muskham on 04.09.1974, a further 45 were found in Nottinghamshire between 1974 and 2007:

Wryneck - birds recorded in Nottinghamshire 1975-2007

Years	1975-79	1980-84	1985-89	1990-94	1995-99	2000-04	2005-07	Total
Birds	16	7	7	2	4	6	3	45

Months	J	F	M	A	M	J	J	A	S	O	N	D
Birds	0	0	0	3	8	0	2	11	19	2	0	0

As can be seen from the above table this species is primarily an autumn migrant with 32 found between 20th August and 3rd October. A further 11 birds were found in the spring between 21st April and 3rd June. Only 5 birds in the period lingered for more than 1 day.
This is an unobtrusive species and several records since 1974 have involved birds which were either found dead (3 birds) or were caught by cats (3 birds). 12 others in the same period were found because they turned up in back gardens.
Extreme dates 08.04.1930 (Rainworth Lodge) - 25.10.1970 (Clumber).

□ **GREEN WOODPECKER** *Picus viridis*
The Green Woodpecker has increased in Nottinghamshire since at least the 1990s and is now a widespread and familiar bird in a variety of habitats such as woods, copses, heathland and suburban parks and gardens. Indications of the current healthy population in the county can be seen from the fact that the species was recorded at 150 sites in 1999 and at 90 sites in the breeding season in 2004.

☐ **GREAT SPOTTED WOODPECKER** *Dendrocopos major*
A common and widely reported species which can be particularly conspicuous when calling in high bare branches, drumming or visiting bird feeders. It is more dependent on woodland or parkland than the Green Woodpecker and may be somewhat scarcer than that species with records at 110 localities in 1999 and 71 sites in the breeding season in 2004. Larger areas of mature woodland often hold several pairs with, for example, 17 territories on the Welbeck estate in 2007.

☐ **LESSER SPOTTED WOODPECKER** *Dendrocopos minor*
In the early 1970s this inconspicuous species was described as scarce and locally distributed in Nottinghamshire with a concentration of records in the western and central parts of the county. This description remains largely true today although the total number of breeding pairs within the county seems to have fallen.
In the three decades from 1975 to 2004 the Lesser Spotted Woodpecker was recorded at almost 200 different localities in Nottinghamshire as set out below:

Lesser Spotted Woodpecker - sites recorded in Nottinghamshire 1975-2004

Years	1975-1984	1985-1994	1995-2004	Total different sites
Birds	87	105	91	190

Although the above table indicates a stable population in the county, the fact that it was recorded at fewer sites in the 10 years from 1995 than in the preceding decade suggests a real decline in numbers given the increased popularity of birdwatching in recent years. What is clear is that birds are now scarce and difficult to locate in Nottinghamshire and the county breeding population may be as low as 25 pairs.
Key sites for the species since 1974 have included Attenborough NR, Wollaton Park, Bestwood CP, Colwick CP, Moorgreen, Sherwood Forest/ Birklands and parts of the Dukeries such as Clumber Park.

☐ **SHORT-TOED LARK** *Calandrella brachydactyla*
There is one old record of this European vagrant which rarely occurs inland - a bird at Nottingham SF on 30.07.1950.

☐ **WOOD LARK** *Lullula arborea*
The Wood Lark has enjoyed mixed fortunes in Nottinghamshire during the last 50 years as it has in other parts of Britain. In the 1950s it was an uncommon and local breeding bird on heathland in the west and north of the county with perhaps as many as 50 pairs breeding. By 1969 the species had virtually vanished as a breeding bird but between 1 and 4 pairs seem to have reappeared 1974-87 (although there are no published records for some years in that period).
There has been a rapid increase in the county population since 1987 with 12 pairs in 1996, 33 territories recorded in 1997 and a peak count of 73

territories in 1998 (with 27 birds using clearfell, 13 on heathland and 16 using brownfield/colliery land). A decade later numbers had fallen back somewhat. Wood Lark bred at 17 sites (centred on Clumber, Welbeck and the greater Sherwood Forest area) with 50-60 singing ♂s in Nottinghamshire in 2005 but only 24 pairs bred in 2006. Despite this Nottinghamshire still holds the strongest population for this species in the Midlands.

Passage movements and wintering are poorly understood but a few migrant birds are recorded each year away from known breeding sites with a concentration of records in February and March and between September and November. Sixteen migrants were found in the Trent Valley 1975-2007. *See NBAR 1997 pp.131-132.*

☐ **SKY LARK** *Alauda arvensis*
A widespread but declining species which is increasingly dependant upon abandoned colliery spoil heaps and other wasteland sites. The population has been reduced by changes in farming over the last few decades and the decline continues to the present date with - for example - records from only 87 sites in 2005 and 56 sites in 2007 compared to 121 sites in 1999. Records published for 2000 indicated a minimum of 411 singing ♂s in the county in that year.

As a winter visitor and passage migrant the species is still reasonably common with some large movements particularly as a result of hard weather:

Sky Lark - Counts of 1000 or more birds in Nottinghamshire 1975-2007

Year	Counts of 1000 or more birds
1979	1000 birds at Barnby Moor 23.01.1979
1985	2000 birds near Eaton Hall 10.01.1985
	1000 birds at Gringley Carr 10.01.1985
1993	1697 birds in the Idle Valley 21.11.1993
1995	1200 birds at Misterton Carr early December 1995
1996	2245 birds over King's Mill Reservoir 31.12.1996
	This is the highest count for 1974-2007.

See NBAR 2003 pp.133-138.

☐ **SHORE LARK** *Eremophila alpestris*
A rare inland visitor to Britain which has occurred in Nottinghamshire 5 times. The first was found at Haywood Oaks near Oxton 31.03.1945 and another was at Attenborough NR 21.10.1973.
7 more birds have been recorded since 1974:
1979   Up to 2 intermittently in the Hoveringham area and at Hazelford Ferry 06.01-04.04.1979.
1986   1 at Lound 25.11-04.12.1986.
1994   4 to the south-west at Colwick CP 03.11.1994.

☐ **SAND MARTIN** *Riparia riparia*
A reasonably common colonial summer visitor which has fluctuated in numbers and is dependent upon the availability of gravel and sand banks for breeding. Numbers were very low in the 1970s but the population recovered over the next 2 decades. In 1995, 18 colonies were reported with a combined total in excess of 1500 nests, a comparable figure to the breeding population in the late 1960s. The gravel pits at Hoveringham have usually held the largest number of breeding birds in recent years with 360 nest holes recorded in 2000. Although numbers had fallen back at this site by 2006, 350 pairs bred at Cottam Power Station in the same year.
Numbers build up at the end of the breeding season and the largest gathering recorded 1974-2007 was of 2000 birds at Attenborough NR 31.07.1976.
The Sand Martin tends to be an early spring migrant.
Extreme dates 28.02.1994 (Lound) - 01.12.1960 (Widmerpool).
See NBAR 1984 pp.44-48.

☐ **BARN SWALLOW** *Hirundo rustica*
A common and widespread summer visitor between April and late September/early October. Large numbers of birds build up in reedbed roosts prior to migration in August and September, particularly at Clumber Park where 10-15,000 were present in August-September 1981, 20,000 on 11.09.1982 and 10,000 on 18.09.1994.
Extreme dates 11.03.1990 (Queen's Medical Centre, Nottingham) - 03.12.1994 (Gringley Carr).

☐ **RED-RUMPED SWALLOW** *Cecropsis daurica*
There were 4 county records of this European vagrant between 1994 and 1996:
1994   Up to 3 birds at Carburton Lodge, Clumber Park 29.10-03.11.1994.
1995   1 at Brinsley Flash on 01.07.1995.
        1 at Attenborough NR on 10.09.1995.
1996   1 at Hardwick Ford, Clumber Park 02.05.1996.
These are the only county records.
See NBAR 1994 p.91 and NBAR 1995 p.116.

☐ **HOUSE MARTIN** *Delichon urbicum*
The House Martin is a common and familiar summer visitor nesting on a wide range of man-made buildings. Birds are normally present between April and October but large post-breeding groups build up in August and September. Such flocks rarely number more than 500 birds but there were 5 counts of 1500 to 2000 birds between 1974 and 2007:

House Martin - counts of 1500 or more birds in Nottinghamshire 1974-2007

Year	Birds recorded
1981	1600 birds at Osberton 19.09.1981
1990	1500 birds at Gunthorpe 05.09.1990
1997	2000 birds at Holme Pierrepont 29.08.1997
1999	1500 birds at Holme Pierrepont 08.08.1999
	c2000 birds at Welbeck 27.08.1999

Extreme dates 18.03.1994 (Colwick) and 18.03.2004 (Sturton-le-Steeple and Torworth) - 30.11.1999 (Bestwood CP).

☐ **RICHARD'S PIPIT** *Anthus richardi*

An eastern vagrant to Britain with 10 Nottinghamshire records. Only 1 was recorded before 1974 - at Misson 24.10.1971.

Ten birds have occurred since then:

1977	1 at Hoveringham and Gibsmere 08-09.10.1977.
1990	1 at Lound 26-29.10.1990.
1994	2 at Holme Pierrepont 07.10.1994.
1996	1 at Lound 16.09-05.11.1996.
	1 at Netherfield 29.09.1996.
2003	1 at Langford Lowfields and Collingham Pits 16.02-29.03.2003.
2005	1 at Annesley Pit Top 26 & 30.09.2005.
2006	1 at Bentinck Void 14.10.2006.
2007	1 at Holme Pierrepont 14.10.2007.

Richard's Pipit is primarily an autumn migrant to Britain and September and October are the peak months for records.
See NBAR 1990 p.58.

☐ **BLYTH'S PIPIT** *Anthus godlewskii*

A first-winter bird at Gringley Carr 28.12.2002-05.01.2003 was an exceptional and well appreciated record of an extremely rare eastern vagrant. This was the twelfth British record.
See NBAR 2002 pp.113-116.

☐ **TREE PIPIT** *Anthus trivialis*

A declining summer visitor to Nottinghamshire. The best evidence of the status of this species comes from a full county survey in 1999 which revealed 394-415 occupied territories in Nottinghamshire with 40% at just 3 sites (Budby Heath, Clipstone Forest and Clumber Park). The survey showed that the range of the species had contracted from 25 occupied 10 km squares 1968-72 to just 9 in 1999 centred on Sherwood Forest district and there must be some concern that the species is on the same downward path as that taken by Common Nightingale, Whinchat and Wood Warbler. Perhaps more encouragingly 54% of territories were in clearfell and young

plantations in 1999, so better woodland management may help this species.
Extreme dates 30.03.1959 (Bramcote) and 30.03.2002 (Fiskerton) - 22.10.1966 (Holme Pierrepont) and 22.10.1993 (Colwick CP).
See NBAR 1999 pp. 145-153.

☐ MEADOW PIPIT *Anthus pratensis*

A common resident breeding bird and passage migrant (March-April and late August-October) in Nottinghamshire with smaller numbers overwintering. Large movements are regular in the autumn with high counts recorded at Annesley (945 south 20.09.1992), King's Mill Reservoir (980 south and south-east 30.09.1996), Eakring (1045 south-west 09.10.2001) and Bentinck Void (1310 on 30.09.2006 - the largest movement 1974-2007). Significant numbers also move through the county in early spring with 1045 east and north-east at King's Mill Reservoir 23-31.03.1996.

☐ WATER PIPIT *Anthus spinoletta*

The Water Pipit is a scarce winter visitor and passage migrant which appears to have become slightly more regular since the mid 1990s. The figures set out below may duplicate returning winter birds:

Water Pipit - birds recorded in Nottinghamshire 1975-2007

Years	1975-79	1980-84	1985-89	1990-94	1995-99	2000-04	2005-07	Total
Birds	15	9	7	8	26	18	4	87
Years	4	4	3	3	5	5	2	26

All records in the period 1974-2007 fell between October and May.
Extreme dates 06.09.1944 (near Netherfield - the first county record) - 18.05.1984 (Attenborough NR).

☐ ROCK PIPIT *Anthus petrosus petrosus*

The Rock Pipit was first recorded in Nottinghamshire as late as 1947 when 2 birds were found at Netherfield 04.04.1947. Only 30 more birds were found between 1947 and 1972. All dated records fell between 24th September and 4th April.

Since 1974 the Rock Pipit has been shown to be a scarce passage migrant and winter visitor to damp habitats with larger numbers of birds identified from the mid 1980s to about 1998. Numbers have fallen back a little since then, which may echo the slight decline in the species as a coastal breeding bird.

Rock Pipit - birds recorded in Nottinghamshire 1975-2007

Months	January-May	June-August	September-December
1975-79	9	0	28
1980-84	2	0	c21
1985-89	7	0	28
1990-94	19	1	c72
1995-99	17	0	44
2000-04	2	0	30
2005-07	3	0	14
Totals	59	1	c237

Records are particularly scarce in May and there is only 1 summer record for the county at Colwick CP 04.06.1994. Otherwise all records fell between 10.09.1993 (Lound) and 18.05.1996 (Hoveringham).
The highest count for any 1 site in the period 1974-2007 is of 5 at Holme Pierrepont 18.10.1998. The county record is of 7 birds at Nottingham SF 05.11.1959.
In addition to the records listed in the table above the **Scandinavian Rock Pipit** *A. p. littoralis* has occurred on at least 7 occasions since 1983:

1983   1 at Holme Pierrepont 29.09.1983.
1995   1 at Girton 22.03.1995.
1997   1 in the Idle Valley 23.03.1997.
2000   1 at Hoveringham 18.03.2000.
2002   1 at Langford Lowfields 17.03.2002.
2003   1 at Holme Pierrepont 28-30.03.2003.
2004   1 at Holme Pierrepont 10.04.2004.

□ YELLOW WAGTAIL *Motacilla flava flavissima*
A reasonably common summer visitor and passage migrant which has declined significantly as a breeding bird in recent years with the loss of damp grassland and small marshes. The stronghold for this species is the Idle Valley where in excess of 55 pairs bred in 1999. Large numbers sometimes appear on migration and 500 at Holme Pierrepont 22.04.1988 is the highest count 1974-2007.
Extreme dates 10.02.1980 (Gunthorpe) - 26.11.1960 (Teversal) with one overwintering at Netherfield and Holme Pierrepont 11.12.1977 - 15.01.1978.
The **Blue-headed Wagtail** *M. f. flava* is a very scarce passage migrant through Nottinghamshire. Most are found in April and May as shown by records for the last 23 years:

Blue-headed Wagtail - birds recorded in Nottinghamshire 1985-2007

Years	1985-89	1990-94	1995-99	2000-04	2005-07	Total
Birds	28	18	35	7	23	111

Months	J	F	M	A	M	J	J	A	S	O	N	D	Undated
Birds	0	0	0	63	37	3	1	4	0	0	0	0	3

1996 was the best recent year for this race with 19 birds including 5 amongst 100 Yellow Wagtails at Holme Pierrepont 08.05.1996.
There are indications that birds may have bred on occasions with a ♂ was seen feeding young at Belmoor in August 1985, another ♂ on territory in the Idle Valley 03.05-15.06.1997 and a ♀ paired with a Yellow Wagtail in the north of the county in 1999.
Birds showing characteristics of other races have been recorded as follows:
Grey-headed Wagtail *M. f. thunbergi* (3 records) - Radcliffe-on-Trent 28.04.1974, Holme Pierrepont 22.04.1988 and Hoveringham 23.05.2004.
Sykes' Wagtail *M. f. beema* (at least 3 records) - Lound 01.05.1990, Girton 11.05.1991 and Cottam PS 02-17.05.1998.

□ **GREY WAGTAIL** *Motacilla cinerea*
The Grey Wagtail was first confirmed as a breeding species in Nottinghamshire in the west of the county in 1955 and may have bred 3 years earlier. This was part of a spread in the breeding range of this species into southern and eastern England. Since then breeding numbers have increased significantly as set out below:

Grey Wagtail - breeding records in Nottinghamshire 1952-1998

Years	1952-1974	1975-1989	1990-1998
Confirmed breeding pairs	9	15	45
Suspected breeding pairs	10	20	49
Maximum pairs in any year	4	6	21

Since 1998 numbers have increased further across the county with a minimum of 28 pairs breeding in 2000.
Otherwise the species is a reasonably widespread winter visitor and passage migrant in small numbers. An unprecedented 31 birds moved south in small parties at Bennerley Marsh 02.10.1995.

□ **PIED WAGTAIL** *Motacilla alba yarrellii*
A common resident breeding species in the county forming substantial winter roosts in urban areas. For example, there were significant roosts at Worksop, Mansfield Railway Station, the Newark by-pass, Retford and Hucknall in the 1990s. Nottingham city centre often holds large numbers of birds in the winter, particularly in and around the Old Market Square and on the traffic island at London Road/Canal Street where 1108 were counted 15.02.1998 - the largest count 1974-2007.
The continental race - the **White Wagtail** *M. a. alba* - is essentially an uncommon passage migrant with small parties occurring between March and May. The largest groups in the period were 10 birds at Hoveringham 16-18.04.1999, 11 at Barton-in-Fabis 19.04.2006, 15 at Colwick CP and Holme Pierrepont 13.04.1985 and 16 at Hoveringham 22.04.2006. Autumn migrants are scarcer. For example, only 14 autumn birds were recorded in the 12 years 1995-2006.

There has been one breeding record for the county involving a White Wagtail - a breeding ♂ trapped at Warsop Vale on 01.06.2005.
Extreme dates 05.03.2000 (King's Mill Reservoir) - 19.11.1995 (Lound).

□ WAXWING *Bombycilla garrulus*
The Waxwing was historically a scarce or uncommon winter visitor to Nottinghamshire (principally between November and March) but larger numbers occurred in invasion years such as 1957-58 and 1965-66.

Birds were scarce in most years 1974-1990 with only c72 birds in total apart from small invasions in 1974 (c37 birds - largest flock c20) and 1975 (33 birds - largest flock 16) and November-December 1988 (43 birds - largest flock 17). 4 years in this period had no records at all and the largest group was just 25 birds at Burntstump Hill crossroads 08.11.1981.

In the period 1991-2007 birds were also scarce in 1994 (3 birds) and 2002 (1 bird) and there were no records in 1993, 1995 and 1998. However, there have been several significant irruptions in this period:

Waxwing - maximum site counts in Nottinghamshire 1990-2007

Winters	1990-91	1991-92	1992-93	1993-94	1994-95	1995-96
Maximum	17	22	0	2	0	510
Winters	1996-97	1997-98	1998-99	1999-00	2000-01	2001-02
Maximum	38	0	1	28	300	0
Winters	2002-03	2003-04	2004-05	2005-06	2006-07	
Maximum	90	90	380	31	12	

Good numbers were present in 1991-92, 1996-97, 1999-2000, 2000-01, early 2003 and 2003-04 and even larger numbers were present in 1996 and 2005 (although birds are always scarce to the east of the A1).

Waxwing - significant invasions into Nottinghamshire in 1996 and 2005

Year	Invasion details
1996	A massive invasion took place between 14.01.1996 and 04.05.1996 peaking at perhaps 1400 birds in the county in early March. Several large and mobile flocks were found, the best being a record 510 at Sherwood in Nottingham 21.02.1996, and a Cedar Waxwing was located amongst the birds in Nottingham.
2005	A slow build up of numbers from November 2004 before huge numbers reached Nottinghamshire in January, February and March 2005 with perhaps 1300 birds in the county in late February. Flocks were more mobile than in 1996, the largest group being 380 at Clipstone Forest 29.01.2005. Unusually, reasonable numbers of birds lingered into early May.

Extreme dates 09.10.2004 (Kirton) - 11.05.2005 (Carrington). There is also an unusual summer record of a bird at Collingham 18-19.08.1971.
See NBAR 1991 pp.51-53.

☐ **CEDAR WAXWING** *Bombycilla cedorum*
The first English record of this species, a first-winter bird, was present with large numbers of Waxwings in central Nottingham (ranging from Wilford in the south to Rise Park in the north) 20.02-18.03.1996 and was widely twitched. The only other British record of this American vagrant is of a bird on the island of Noss, Shetland on 25-26.06.1985.
See NBAR 1996 pp.86-88 and Birding World Vol. 9/2 (February 1996) pp.70-73.

☐ **DIPPER** *Cinclus cinclus gularis*
Although the Dipper breeds in nearby Derbyshire, it is a rare visitor to Nottinghamshire with only 18 birds in the 34 years since 1974.

Dipper - birds recorded in Nottinghamshire 1975-2007

Years	1975-79	1980-84	1985-89	1990-94	1995-99	2000-04	2005-07	Total
Birds	1	3	3	1	2	8	0	18

Months	J	F	M	A	M	J	J	A	S	O	N	D
Birds	1	3	2	0	0	1	1	1	2	4	2	1

15 of the 18 records since 1974 occurred in the period 4th September to 14th March but there were 2 summer records - at Hoveringham 27.06.1984 and at Ollerton 05.07.1995. The other unseasonal record was of a long staying bird at Shireoaks on the Chesterfield Canal 09.08.2000 - 30.01.2001.
There are just 8 records of 11 birds before 1974 including a single continental **Black-bellied Dipper** *C.c. cinclus* shot on the River Greet at Southwell in c1873. All 11 birds were found between November and January.

☐ **WREN** *Troglodytes troglodytes*
The Wren is a very common resident throughout Nottinghamshire with birds present in a wide variety of habitats. This species suffers from high mortality in very cold winters, for example in 1963 (when perhaps 90% of the population was wiped out) and in 1978-79. However, numbers tend to recover quickly. An indication of the size of the population at its peak can be seen from counts of 158 territories at Attenborough NR in 2003 and 121 territories at Centre Parcs in 2004.

☐ **DUNNOCK** *Prunella modularis*
The Dunnock is another common but under-recorded resident species throughout Nottinghamshire. In the early 1970s breeding densities of perhaps 8-12 pairs per 100 acres on farmland and 25-30 pairs per 100 acres in woodland were estimated for the county. Thirty years later the species is still present in almost any habitat with a few shrubs available for breeding. An indication of the size of the population can be seen from counts of 60 singing ♂s at Attenborough NR, 33 singing ♂s at Bestwood

CP and 28 singing ♂s at Eakring Meadows in 1998 and 56 breeding pairs at Centre Parcs in 2003.

ROBIN *Erithacus rubecula melophilus*
A common resident species with woodland breeding densities recorded at 40 pairs per 100 acres in the early 1970s. An indication of the size of the population can be seen by counts of 78 territories at Clumber Park in 1999, 71 territories at Bestwood CP the same year, 87 at Treswell Wood in 2003 and 138 territories at Centre Parcs in 2002.

Small autumn and winter influxes - of what are presumably **Continental Robins** *E.r.rubecula* - are recorded on occasions with 60 birds at Colwick CP 28.09.1997, 79 through Lound on 06.10.2002 and 65 at Holme Pierrepont 16.01.2005.

COMMON NIGHTINGALE *Luscinia megarhynchos*
Nottinghamshire is at the northern edge of the range of Common Nightingale and the species is just hanging on as a migrant breeding bird in the county. A survey of the species in 1911-12 showed it to be a common bird in the Trent Valley (between Nottingham and Newark) and around Southwell with breeding birds in the Vale of Belvoir and in the Dukeries. At its peak there were perhaps 50-100 pairs in the county concentrated in the Dukeries.

Sadly the story from 1974-2007 has been one of almost continuous decline from 14 singing birds in a full survey in 1976 to half that number in a second survey 4 years later. Up to 9 birds were recorded singing annually between 1981 and 1996 but the population collapsed thereafter with only 3 records of single singing birds 1997-2003. Since then birds have recolonised a site in the north of the county from neighbouring breeding sites in Lincolnshire, with 3 birds in song in 2004 and 4 birds in 2005 and 2006. Breeding was finally proved at this site in 2006 and up to 9 ♂s were present in 2007.

Occasional migrants still turn up well away from known breeding sites. Extreme dates 10.04.1961 (near Newark) - 31.08.1976 (Cottam).

BLUETHROAT *Luscinia svecica*
There were 1 or probably 2 county records of this species.

In 1979 a ♂ of the **white-spotted race** *L. s. cyanecula* held territory at Tiln from early April to 12.05.1979. The same or perhaps another bird was trapped at Attenborough NR 09.06.1979.

☐ BLACK REDSTART *Phoenicurus ochruros*

This species first bred in Nottinghamshire in Nottingham city centre in 1958 and has remained a rare breeding bird in the county. In the 1970s up to 5-7 ♂s held territories in the city and 2-5 ♂s sang in Nottingham from the mid-1980s through to 1999. However the redevelopment of old buildings in the Lace Market and other parts of the city centre has removed many potential nesting sites for this species and a pair which bred successfully in 2003 is the only recent published record. Breeding has never been recorded elsewhere in Nottinghamshire.

Occasional migrants and winter birds turn up elsewhere in the county:

Black Redstart - birds recorded in Nottinghamshire (excluding birds in Nottingham city centre) 1975-2007

Years	1975-79	1980-84	1985-89	1990-94	1995-99	2000-04	2005-07	Total
Birds	12	9	10	12	19	16	2	80

Months	J	F	M	A	M	J	J	A	S	O	N	D	Undated
Birds	2	5	9	17	6	4	2	1	2	17	11	3	1

Three birds have overwintered in the county in recent years - 1 at Lound 13.11.1986-10.01.1987, a ♀ at Netherfield 23.11.1997-February 1998 and a ♀ in Nottingham 22.12.2006-06.03.2007.

◻ COMMON REDSTART *Phoenicurus phoenicurus*
The Common Redstart is a summer breeding bird in old woods in the western half of the county, particularly in the Dukeries and in the Sherwood Forest - Birklands area. However it has declined in Nottinghamshire since the 1970s and many breeding sites have been lost as set out below:

Common Redstart - breeding pairs in Nottinghamshire 1950-1996

Decade	Sites	Estimated pairs
1950-59	15	96
1960-69	32	94
1970-79	15	81
1980-89	9	65
1990-96	4	14

After 1996 there was something of a recovery with as many as 32-37 ♂s in the county in 2002. However the general picture is still worrying with only 3 ♂ located in 2006 and an increased concentration on a few traditional sites at Budby, Clumber Park, Clipstone Forest and particularly the Birklands/Sherwood Forest area.

Extreme dates 27.03.1967 (Burton Joyce) - 26.11.1970 (locality unrecorded).

◻ WHINCHAT *Saxicola rubetra*
Formerly a fairly common summer visitor, this species has virtually disappeared as a breeding bird in Nottinghamshire since 1974, falling from 34 pairs in 1974 (and even 22-23 territories at 9 sites in 1980) to no pairs at all in 1995. The only subsequent breeding records are of single pairs which bred at Gringley Carr in 2002 and at Misterton Carr in 2007. A pair also summered at Budby in 1997.

Whinchat - breeding numbers in Nottinghamshire 1950-2007

Decade	Sites	Estimated pairs
1950-59	37	85
1960-69	25	48
1970-79	32	80
1980-89	27	56
1990-99	3	5
2000-07	2	2

This decline has occurred in other lowland English counties and is possibly linked to agricultural intensification which limits breeding and feeding opportunities for this insectivorous species. However many pairs formerly bred in young conifer plantations and on heathland in the county where substantial breeding habitat remains. This suggests that the collapse of the breeding population has resulted from complex climate changes or other factors which are as yet poorly understood.

The Whinchat remains a fairly common spring and autumn migrant through the county.

Extreme dates 06.04.1980 (Lound) - 21.11.2000 (Brierley Forest CP). In addition an exceptionally early migrant or winterer was at Ratcliffe-on-Soar 22.02.1959.

☐ COMMON STONECHAT *Saxicola torquata*

The Common Stonechat once bred in Nottinghamshire in small numbers, for example around Mansfield. Today it is an occasional breeding species and available habitat for this species is now scarce in the county. There are only 4 published breeding records for the period 1974 - 2007 - at Budby Common in 2001 and 2006 and at Clumber Park in 2003 and 2004.

The recent resumption of breeding may be part of a general upturn in the fortunes of this species which has seen wintering numbers increase from as few as 2-9 birds per year between 1980 and 1987 to regular double-figure counts of wintering birds from the late 1990s onwards as set out below:

Common Stonechat - peak November and December numbers in Nottinghamshire 1997-2007

Years	97	98	99	00	01	02	03	04	05	06	07
Birds	13	17	21	55	68	54	26	23	68	64	61

☐ NORTHERN WHEATEAR *Oenanthe oenanthe oenanthe*

The Northern Wheatear is a fairly common passage migrant through the county, arriving in late March and continuing in good numbers through to May each year. The largest group of spring migrants is of at least 30 birds at Netherfield on 23.04.2006 and there were 24 birds in the Idle Valley 26.04.1997. Returning birds occur in smaller numbers in August and September.

The species formerly bred in small numbers at localities such as Ratcher Hill and Oxton Warren but there is only one published breeding record since 1948, in the north-east of the county in 1970. Birds may also have bred at Daneshill in the mid 1980s and there is also a recent intriguing record of a recently fledged juvenile at Misson Carr in June 2004.

Extreme dates 26.02.2002 (Attenborough NR) - 13.11.1989 (Langley Mill).

The **Greenland race** *O. o. leucorhoa* is a passage migrant in small numbers, although birds may be under-recorded.

Greenland Wheatear - birds recorded in Nottinghamshire 1975-2007

Years	1975-79	1980-84	1985-89	1990-94	1995-99	2000-04	2005-07	Total
Birds	1	2	6	0	14	18	c52	c93

Months	J	F	M	A	M	J	J	A	S	O	N	D
Birds	0	0	0	28	c44	0	0	5	13	3	0	0

RING OUZEL *Turdus torquatus*

The Ring Ouzel is a regular migrant through Nottinghamshire in very small numbers with 1-17 birds recorded in 28 years between 1974 and 2007.

Ring Ouzel - birds recorded in Nottinghamshire 1975-2007

Years	1975-79	1980-84	1985-89	1990-94	1995-99	2000-04	2005-07	Total
Birds	6	6	24	15	34	23	44	152

Months	J	F	M	A	M	J	J	A	S	O	N	D
Birds	0	1	8	93	12	0	0	0	6	25	7	0

In total 152 birds were recorded in the period with 113 spring records (18th March - 28th May) and 38 autumn records (6th September - 26th November). The largest group reported on migration was 3♂ and 3♀ at Toot Hill near Kneeton on 17.04.1987 and as many as 10 passed through Annesley Pit Top in April and May 2006.

In addition to the migrant birds there was one winter record in the period - a ♂ in Ilkeston Road, Nottingham 03.02-27.03.1996. There are 5 previous winter (December-February) records for the county between 1865 and 1963.

The species apparently bred in Sherwood Forest in 1856 but there are no subsequent breeding records.

□ BLACKBIRD *Turdus merula*
A very common and familiar resident, passage migrant and winter visitor. Sample census counts of 140 nesting pairs at Centre Parcs in 2002 and 105 pairs at Eakring in 2000 indicate that there are large numbers breeding throughout the county.

Large roosts have also been recorded in the county on occasions although such gatherings have been less frequent in recent years. This may be a result of under-reporting or perhaps represents a genuine decline in the wintering population in the county.

Blackbird - maximum site counts in Nottinghamshire 1975-2007

Years	1975-79	1980-84	1985-89	1990-94	1995-99	2000-04	2005-07
Site maxima	1000	750	480	70	137	210	108

The largest site counts since 1974 involve 1000 birds roosting at Attenborough NR on 11.11.1978 and 10.11.1979.

□ DUSKY THRUSH *Turdus naumanni eunomus*
One old record - the first for Britain - a ♂ shot near Gunthorpe on 13.10.1905.
There have only been 10 records of this species in Britain.

□ FIELDFARE *Turdus pilaris*
A common winter visitor from late September to April with a few stragglers to early May. Large numbers often move through the county ahead of hard weather. Flocks of over 1000 birds are not uncommon most years but there were massive winter numbers in the Idle Valley in the early to mid 1990s with a high of 12840 there on 23.10.1993.

By contrast summer records (June-August) are extremely unusual and breeding has never been proved. However there is an intriguing record of a bird carrying food at an undisclosed locality 01.05.1984. Fieldfares bred irregularly in nearby Staffordshire and Derbyshire on several occasions 1969-1996.

☐ SONG THRUSH *Turdus philomelos clarkei*

The Song Thrush is a common resident species in a wide range of habitats (such as woodland, parks and gardens) which has undergone a significant decline in breeding numbers since the late 1980s. However numbers may have stabilised recently with records from 59 to 121 sites annually since 1998 and some sites still hold many pairs with - for example - 36 to 60 pairs breeding annually at Centre Parcs between 1999 and 2007. In addition to the resident population within Nottinghamshire there are occasional autumn and winter influxes (such as 50 birds at Hoveringham 20.11.1983) which involve either continental migrants *T.p.philomelos* or perhaps birds displaced from other parts of Britain.

☐ REDWING *Turdus iliacus*

The Redwing is a common winter visitor with the first migrants arriving in September and the last few birds leaving in May. Large flocks often feed in hedgerows and on farmland and 4000 birds at Bennerley Marsh on 05.02.1996 is the largest such gathering 1974-2007. The run of mild winters in recent years has seen fewer birds wintering in the county. However large numbers still move through the county on passage, with as many as 4290 south-west over Annesley Pit Top on 26.10.2006 and a further 4950 south-west over the same site the following day.

Summer records (June to August) are very unusual but single birds have been recorded in all 3 summer months - at Ashfield Pit Top 18-20.06.1998, at Mapperley in Nottingham 24.07.1984 and at Collingham 10-12.08.1990.

☐ MISTLE THRUSH *Turdus viscivorus*

This species is a reasonably common and widely distributed resident species which breeds early in the year. As a result Mistle Thrushes tend to form large post-breeding flocks between June and October. 137 at Oldcotes 17.08.1998 was the largest such group between 1974 and 2007.

☐ CETTI'S WARBLER *Cettia cetti*

There are 8 records of this species - which colonised parts of southern Britain in the 1960s and 1970s- including 4 records in 2007:

1981	1 at Holme Pierrepont 15-27.03.1981.
1994	1 at Cottam on 30.10.1994 (which had been ringed at Brandon Marsh, Warwickshire on 02.10.1994).
2004	An intriguing record of 1 or perhaps 2 birds at Misson Carr 12-13.05 & 29.06.2004 with 1 bird seen carrying food on the last date.
2005	1 at Attenborough NR on 24.03.2005.
2007	1 at Attenborough NR from 11.03.2007. A second bird was present from April and the pair raised 4 young.
	1 at Cottam PS 04.11.2007
	1 at Clifton Grove 28.11-09.12.2007.
	1 at Netherfield 16.12.2007.

This is a species which now breeds in a number of small Midlands reserves in Northamptonshire, Warwickshire and Worcestershire and appears to be on the verge of becoming a regular breeder in Nottinghamshire.

☐ **GRASSHOPPER WARBLER** *Locustella naevia*
The Grasshopper Warbler is an uncommon and secretive summer migrant which is normally detected by its unusual reeling call. Breeding birds favour damp, tangled scrub with low bushes at sites such as Netherfield (where there were 11 reeling ♂s in April 2007). It has always been a local species but there are some recent indications that the species may be declining in Nottinghamshire and a full survey of all likely sites would be welcome.

Grasshopper Warbler - sites in Nottinghamshire 1999-2005

Years	1999	2000	2001	2002	2003	2004	2005
Sites	32	26	17	22	21	22	40

Extreme dates 30.03.1989 (Colwick CP & Edwinstowe) - 01.10.2006 (Netherfield).

☐ **SAVI'S WARBLER** *Locustella luscinioides*
This very rare British breeding bird was recorded in Nottinghamshire 3 times in the 1990s:
1990   1 in song at Besthorpe 18.05-c.07.07.1990.
1996   1 in song at Attenborough NR 09-12.06.1996.
1997   1 in song at Attenborough NR 10-11.05.1997.
These are the only county records.
See NBAR 1990 p.57.

☐ **SEDGE WARBLER** *Acrocephalus schoenobaenus*
The Sedge Warbler is a common summer visitor. Large numbers of singing ♂s are present at most damp reedy sites in the Trent Valley and elsewhere. For example Attenborough NR held 118 territories in 1979, Holme Pierrepont held 54 ♂s in 1999, Netherfield had 60 ♂s in 2004 and Langford Lowfields had 60 pairs in 2006.
Extreme dates 31.03.1998 (Netherfield) - 07.11.1979 (Mansfield).

☐ **MARSH WARBLER** *Acrocephalus palustris*
A very rare summer visitor to Nottinghamshire with just 7 records. The first record involved a breeding pair at Attenborough NR in 1969 and another was in song there 01-02.07.1972.
Subsequently 5 singles have been found:
1986   1 in song at Cottam 01.06.1986.
1997   1 in song at Attenborough NR 21-22.05.1997.
         1 in song at Netherfield 14-15.06.1997.
2000   1 in song at Danethorpe 03.06.2000.
2004   1 at Mill Lakes, Bestwood 26.06.2004.
See NBAR 1997 pp.140-142.

☐ **REED WARBLER** *Acrocephalus scirpaceus*
This species is a fairly common summer visitor breeding in all major reedbeds in the Trent Valley and elsewhere (for example at Attenborough NR, Holme Pierrepont, Netherfield and Hoveringham in the Trent Valley and Lound in the Idle Valley). Some of these sites hold large numbers with 150 pairs at Attenborough NR in 1981 and 70 singing at Holme Pierrepont in 2004.
Extreme dates 31.03.2005 (Holme Pierrepont) - 03.11.2002 (Holme Pierrrepont).
See NBAR 2005 pp.107-108.

☐ **GREAT REED WARBLER** *Acrocephalus arundinaceus*
There are two fully acceptable records of this chunky warbler:
1976   1 at Bulcote 19.08.1976.
1997   1 at Netherfield 12-29.06.1997.
Three quarters of all English records involve birds found in May and June.
See NBAR 1997 pp.140-142.

☐ **ICTERINE WARBLER** *Hippolais icterina*
There is one county record of a bird in song at Colwick 13.07.1945. This species is a scarce coastal migrant to Britain and inland records are extremely unusual.

☐ **BLACKCAP** *Sylvia atricapilla*
A common summer visitor which outnumbers the closely related Garden Warbler at most sites. Examples of the strength of the local population include a count of 75 territories at Attenborough NR in 2002 and 38 at Welbeck in 2007. It is more adaptable than some warbler species and odd pairs even breed near to the city centre in Nottingham and around other towns. The Blackcap has also become a regular overwintering bird in the county since the 1950s and the winter population has grown significantly from the early 1970s with many records from gardens.

☐ **GARDEN WARBLER** *Sylvia borin*
This warbler is a fairly common summer visitor to Nottinghamshire with significant population counts from some of the well watched localities in the Trent Valley. For example there were estimates of 31 territories at Attenborough NR in 2002 and 16 territories at Holme Pierrepont in 2004.
Extreme dates 29.03.1969 (Clumber) - 25.10.1998 (Treswell Wood).
In addition there are 5 late November-December records between 1961 and 1991 - the latest 2 birds at Attenborough NR on 20.12.1969.

☐ **BARRED WARBLER** *Sylvia nisoria*
There are 4 August -September records of what is primarily an East Coast migrant to Britain. The first was at Radcliffe-on-Trent 08.09.1968.
Three further birds have been recorded since 1974:
1977   1 at Annesley Woodhouse 20.09.1977.
1990   1 at Hills and Holes 23.09.1990.
1999   A juvenile trapped on the Welbeck Estate 04.08.1999.

☐ **LESSER WHITETHROAT** *Sylvia curruca*
A fairly common summer visitor to many parts of the county but quite often under-recorded. The only significant population estimates for 1974-2007 are modest (and perhaps misleading) counts of 37 territories in 26 areas in 1980 and 51 singing ♂s in 2002. What these counts do show is that birds are generally not found in large numbers anywhere but are thinly distributed throughout the county.
Extreme dates 05.04.1994 (Attenborough NR) - 23.10.1994 (Awsworth). However there are 3 November records (the last at Gedling Pit Top 23-24.11.1997), 1 December record (hit by a car at Bingham 08.12.1978) and 1 January record (Blyth 13.01.1979) between 1974 and 2007. Another was present in a West Bridgford garden 11.02-01.03.1996.

☐ **COMMON WHITETHROAT** *Sylvia communis*
A common summer visitor which experienced sudden declines in the late 1960s and 1980s but has since recovered and is now widespread as a hedgerow and scrubland breeding bird. Two countywide counts within the period 1974-2007 illustrate the changing fortunes of this species - a low of just 39 territories in 1983 and a high of 345 singing ♂s in 1999. However numbers may have fallen back a little since 1999.
The Idle Valley is a particular stronghold for this species with an estimate of 150 singing ♂s in 1993.
Extreme dates 22.03.2002 (Lound) - 13.10.1991 (Holme Pierrepont) with possible overwintering birds at King's Mill Reservoir on 28.11.2004 and at Attenborough NR on 10.12.1966 and 22.01.1972.

☐ **PALLAS'S LEAF WARBLER** *Phylloscopus proregulus*
A bird seen by 1 lucky observer at Bolham Hall on 25.11.2002 was the first county record of this eastern vagrant. Most British records are in October and November.
See NBAR 2002 p.119.

☐ **YELLOW-BROWED WARBLER** *Phylloscopus inornatus*
Five examples of this classic East Coast autumn migrant have reached Nottinghamshire:
1994   1 at Colwick CP 30.09.1994.
       1 at Treswell Wood 30.10.1994.

2004   1 at Colwick CP 21.12.2003-07.01.2004.
        1 at Belmoor near Lound 29.09.2004.
2006   1 near Clumber Park 08.10.2006.
An earlier record of 2 birds at Lowdham 18.10.1914 is no longer considered acceptable.
See NBAR 1994 p.92.

☐ WOOD WARBLER *Phylloscopus sibilatrix*

The Wood Warbler has always been an uncommon summer visitor to Nottinghamshire, particularly to the old woods in the Birklands - Sherwood Forest area. In the past some sites held significant numbers of breeding birds and Joseph Whitaker recorded about 30 pairs nesting in Harlow Wood at the turn of the twentieth century.

However the period 1974-2007 has witnessed the virtual disappearance of this species from the county:

Wood Warbler - maximum territories in Nottinghamshire 1975-2007

Years	1975-79	1980-84	1985-89	1990-94	1995-99	2000-04	2005-07
Maximum territories	7	16	12	8	2	0-1	0-1

The best recent year was 1984 with 16 territories but only 2 areas held birds by 1992. After 1992 breeding was only proved in 1996 (at Church Warsop and Clumber) and the only recent evidence of breeding is of single birds in song in 1999, 2000, 2001 and 2006.

Such has been the recent decline in the fortunes of this species that in the period 1997-2007 only 25 other migrant birds were found in the county.

Wood Warbler - migrant birds in Nottinghamshire 1997-2007

Months	J	F	M	A	M	J	J	A	S	O	N	D
Birds	0	0	0	10	11	1	0	2	1	0	0	0

Extreme dates 03.04.1886 (locality unrecorded) - 18.09.1960 (Clumber).

☐ COMMON CHIFFCHAFF *Phylloscopus collybita collybita*

The Common Chiffchaff is a common summer visitor and regular winter visitor in small numbers. As a breeding bird this species may now be doing better than the closely related Willow Warbler in some places. For example there were twice as many breeding Willow Warblers than Common Chiffchaff at Centre Parcs in 2000. However by 2004 there were 33 pairs of Common Chiffchaff there, easily outnumbering the 14 pairs of Willow Warblers present at the site.

At the same time the number of wintering birds has increased steadily since approximately 1975 and overwintering has been annual since 1992.

Siberian or **Scandinavian Chiffchaffs** with characteristics of 1 of the Eastern races *P.c.tristis* or *P.c.abietinus* have been claimed on at least 9 occasions:

1986	1 at Attenborough NR 22.11.1986.
1991	1 at Hills and Holes 01.12.1991.
1995	1 at Clifton Grove 17.01.1995.
1996	1 at Shireoaks 27-28.09.1996.
2000	1 at Nottingham General Cemetery 27.08.2000.
	1 at King's Mill Reservoir 13.11.2000.
	1 at Misson Carr 06.12.2000.
	1 at Calverton SF end December 2000.
2002	1 at Colwick CP 12-28.12.2002.

☐ **WILLOW WARBLER** *Phylloscopus trochilus trochilus*

The Willow Warbler is a common summer visitor to many parts of the county, breeding in woodland, copses, parkland and overgrown wasteland. However this species has perhaps not fared as well as Common Chiffchaff at some sites recently. For example numbers at Centre Parcs have fallen from 102 territories in 1994 to 14 pairs in 2004.

Strong numbers occur on passage - particularly between late March and May - when significant site counts are sometimes recorded. For example 120 birds were recorded at Bennerley Marsh on 15.04.1995.

Extreme dates 07.03.1997 (Coxmoor GC) - 12.11.1986 (Colwick CP). Exceptionally 1 wintered at Misson 30.01-15.03.2005.

Birds showing characteristics of the **Northern** or **Scandinavian** form *P.t.acredula* were noted at Oxton Bogs in the autumn of 1983 and at Clumber Park 28.05.2000.

☐ **DUSKY WARBLER** *Phylloscopus fuscata*

The first county record of this eastern vagrant involved a bird at Netherfield Lagoons on 07.10.2007.

See NBAR 2007 pp.121-124.

☐ **GOLDCREST** *Regulus regulus*

This is a common resident breeding bird (particularly in coniferous woodland) throughout Nottinghamshire. Outside the breeding season birds join winter tit flocks and large groups are sometimes recorded at this time.

Goldcrest - counts of 100 or more birds in Nottinghamshire 1975-2007

Year	Birds recorded
1977	100 birds at Ollerton late October 1977
1985	c200 birds at Stapleford Woods 29.10.1985
1988	Over 200 birds at Colwick CP 19.10.1988
1998	100 birds at Walesby Common 06.11.1998
2001	100 birds at Budby Common 20.01.2001
2003	100 birds at Sherwood Forest 17.02.2003

☐ **FIRECREST** *Regulus ignicapilla*
The Firecrest was formerly an extremely rare visitor to Nottinghamshire with just 8 records between 1850 and 1972. All dated records fell between 1st December and 25th March except for a bird at Thorpes Wood, Coddington on 01.05.1958.
Since 1974 it has occurred far more regularly - particularly in the winter - with only 2 blank years (1976 and 1984).
Following a bird which overwintered at Attenborough NR from 01.12.1974 to 02.02.1975, there were a further 97 records in the county to 2007:

Firecrest - birds recorded in Nottinghamshire 1975-2007

Years	1975-79	1980-84	1985-89	1990-94	1995-99	2000-04	2005-07	Total
Birds	10	5	17	13	17	24	11	97

Months	J	F	M	A	M	J	J	A	S	O	N	D
Birds	12	11	10	3	4	1	2	2	1	7	29	15

Clumber Park is easily the best locality for this species with 22 records in the period and up to 4 birds wintered together at Colwick CP between November 2002 and February 2003.
See NBAR 1985 pp.51-53.

☐ **SPOTTED FLYCATCHER** *Muscicapa striata*
The Spotted Flycatcher was formerly a reasonably common summer visitor to Nottinghamshire but numbers declined sharply in the 1980s and 1990s, as in other parts of Britain. A full survey carried out in 2004 revealed just 124 territories at 59 sites and should provide a baseline for later research.
Extreme dates 18.04.1977 (Norwell) - 14.10.1981 (East Bridgford) and 14.10.1983 (locality unrecorded).
See NBAR 2004 pp.120-125.

☐ **RED-BREASTED FLYCATCHER** *Ficedula parva*
This East Coast migrant has straggled inland to Nottinghamshire on 3 occasions with two birds before 1974 - a ♀ at Nottingham SF 26.08.1947 and a ♀ nearby at Burton Meadows on 13.08.1950.
One has occurred since:
1993   1 trapped at Treswell Wood 24.10.1993.

☐ **PIED FLYCATCHER** *Ficedula hypoleuca*
The Pied Flycatcher is primarily a very scarce migrant through Nottinghamshire with a handful of birds recorded virtually annually.
125 migrants passed through the county from 1975:

Pied Flycatcher - migrant birds in Nottinghamshire 1975-2007*

Months	J	F	M	A	M	J	J	A	S	O	N	D	Undated
Birds	0	0	0	32	38	1	3	27	20	3	0	0	1

* The table excludes breeding records, territorial ♂s and prospecting pairs.

1996 was the best recent year with 17 records including 4 singing ♂s. Nottingham General Cemetery has been a particularly good site for migrant birds in recent years with 2 spring and 11 autumn records 1996-2007.
There are also a handful of breeding records for the county. Joseph Whitaker suggested that the species bred twice in the county before 1907 (at Ramsdale in 1875 and at Birklands in an unrecorded year) and a pair apparently nested between Daybrook and Mapperley in 1923. More recently there were 1-4 singing ♂s or prospecting pairs in 8 years between 1974 and 2001. Most of these records have been at Clumber Park and nearby sites. More significantly a pair bred unsuccessfully at Clumber in 1986 and a pair probably bred there in 1989. In 1998 a pair bred successfully at an unnamed site in the county. However, breeding activity seems to have tailed off in the last decade.
Extreme dates 02.04.1960 (Fountaindale) - 06.11.1971 (Girton).

□ BEARDED TIT *Panurus biarmicus*
The Bearded Tit is a rare and irregular late autumn visitor to reedbed sites in Nottinghamshire. The first county record was one at Gedling in 1922. Subsequently there were 3 records from Attenborough NR - up to 9 birds 20.11.1965-06.02.1966, up to 5 birds 20.10-23.11.1968 and up to 3 birds 12-19.11.1972 - and 5 birds were at Holme Pierrepont on 23.10.1968.
Following 2 further birds at Attenborough NR 20.01-17.03.1974 there were records in 15 years between 1975 and 2007 involving at least 74 more birds. 25 of these birds were recorded in the autumn of 1977 when there was a significant influx of birds into the county. This influx included a record count of 14 at Holme Pierrepont on 02.10.1977.

Bearded Tit - birds recorded in Nottinghamshire 1975-2007

Years	1975-79	1980-84	1985-89	1990-94	1995-99	2000-04	2005-07	Total
Birds	37	12	2	2	1	11	9	74

Months	J	F	M	A	M	J	J	A	S	O	N	D
Birds	1	2	2	2	0	1	0	2	0	53	7	4

The extensive reedbeds at Attenborough NR are the favoured site for this species with 18 birds from 1974. Holme Pierrepont had 18 birds in the same period (including the 14 in 1977) and 8 have been found at Netherfield since 2001. In contrast only 8 birds have ever been recorded away from the Trent Valley including 6 birds at King's Mill Reservoir since 1974.
Birds ringed at Blacktoft Sands (1977) and in Suffolk (1968) have been found in the county indicating that most records involve post-breeding irruptive movements from Britain's localised breeding population.

☐ **LONG-TAILED TIT** *Aegithalos caudatus*
A reasonably common resident species throughout the county. Long-tailed Tits suffer in hard winters when mortality can be high but with the recent run of mild winters there are currently healthy numbers in Nottinghamshire as shown by three figure site counts. There were 9 published site counts of more than 100 birds in Nottinghamshire 1974-2007 - 8 of these counts have been made since 1995 including the highest county site count of 168 birds at Lound 06.09.2002 and the largest single flock of 150 at Budby South Forest 09.02.2003.

☐ **MARSH TIT** *Poecile palustris*
A fairly common resident species which is outnumbered by the closely related Willow Tit in Nottinghamshire. The Marsh Tit generally prefers more mature woodlands such as Clumber Park and Welbeck and therefore has a more limited distribution in the county with records at 30 to 59 sites per year 1998-2007. 20 at Clumber Park on 30.11.2003 is the largest site count recorded since 1974.

☐ **WILLOW TIT** *Poecile montana*
A fairly common resident bird which outnumbers the Marsh Tit in Nottinghamshire. The county is currently regarded as a national stronghold for this species as there is plenty of damp woodland available for nesting birds. As a result, the species is reasonably widely distributed and was recorded at 56 to 103 sites per year between 1998 and 2007. 30 birds in the Sutton-cum-Lound/Lound GP complex on 01.08.2005 and 17 at Ranskill on 26.07.2005 are the largest published site counts between 1974 and 2007.
See NBAR 2007 pp.112-120.

☐ **COAL TIT** *Periparus ater britannicus*
A reasonably common resident, particularly in the coniferous and mixed woods in the heart of the county with - for example - 65 pairs breeding at Centre Parcs near Rufford in 2001. 70 at nearby Budby on 02.01.1988 is the largest published count for 1974-2007.
1 or 2 birds of the **Continental European race** *P.a.ater* were reported from Nottingham General Cemetery between September and November 1997 and another was at Annesley Pit Top on 04.12.2005.

☐ **BLUE TIT** *Cyanistes caeruleus*
A common and widespread woodland resident throughout Nottinghamshire. Birds often live close to humans with numerous birds taking advantage of nest boxes and using garden bird feeders. 93 birds at Holme Pierrepont on 13.02.2000 and 95 there on 16.01.2005 are the largest published counts for 1974-2007.

☐ **GREAT TIT** *Parus major*
A common and widespread resident species occupying the same habitats as the Blue Tit. Sample census counts of breeding birds indicate the strength of the population across the county with - for example - 72 pairs in nestboxes at Birklands in 1985, 50 pairs at Colwick CP in 1999 and counts of 58 pairs breeding in Treswell Wood in 2003 and 2004.
Small numbers join autumn tit flocks but very large post-breeding groups are unusual. The largest count in the period 1974-2007 was 60 birds at Clumber Park on 07.07.1994.

☐ **EUROPEAN NUTHATCH** *Sitta europaea*
A reasonably common resident of woodland and parkland in the county but more local than the Eurasian Treecreeper which occupies the same habitat. Prior to 1974 the species was recorded as having a predominantly western distribution within the county with the stronghold for the species at Birklands and in the Dukeries.
Thirty years later the major stronghold remains centred upon Sherwood Forest - Rufford - Centre Parcs and the major woodland and parkland of the Dukeries (such as Clumber Park and Welbeck). Some of these sites hold large numbers of breeding birds with - for example - 34 territories recorded at Welbeck in 2003. However the species is also present at urban and suburban sites such as Wollaton Park, the Nottingham University campus and Nottingham General Cemetery and may be increasing.

☐ **EURASIAN TREECREEPER** *Certhia familiaris*
The Eurasian Treecreeper is an unobtrusive but reasonably common resident of woodlands throughout the county. The larger tracts of mature woodland and parkland in the county all support several breeding pairs. For example 16 pairs were recorded at Welbeck in 2002 and 18 to 21 pairs were recorded as breeding at Centre Parcs 2000-2007.
Birds are often more obvious during the autumn and winter months when small numbers often travel with tit flocks between different blocks of woodland.

☐ **PENDULINE TIT** *Remiz pendulinus*
A bird which showed briefly at Attenborough NR on 23.10.1994 is the only county record of this European vagrant. There were 210 records of Penduline Tit in Britain between 1966 and 2007.
See NBAR 1994 p.93.

☐ **GOLDEN ORIOLE** *Oriolus oriolus*
The Golden Oriole is a true rarity in Nottinghamshire with 8 records between 1863-1973 including 1 pair near East Leake in 1917 and single pairs at Oxton and at Stanford-on-Soar in 1947.
Since 1974 another 15 to 17 birds have been recorded:

Golden Oriole - birds recorded in Nottinghamshire 1975-2007

Years	1975-79	1980-84	1985-89	1990-94	1995-99	2000-04	2005-07	Total
Birds	2	0	4	4-5	3-4	1	1	15-17

Months	J	F	M	A	M	J	J	A	S	O	N	D
Birds	0	0	0	0	6	7-9	2	0	0	0	0	0

1977   1 in song at Clumber Park 28.05.1977.
1 near Thorney 14.06.1977.
1988   A 1st summer ♂ sang in an unnamed wood in north Nottinghamshire 04-05.06.1988.
1989   A ♀ near Worksop 13.05.1989.
A ♀ at Peafield Plantation near Warsop 17.07.1989 with a ♂ there 19.07.1989.
1992   A ♂ in song at Arnold 23.05.1992.
A ♂ in song at Holme Pierrepont 25.05.1992.
A ♂ in song (and a possible ♀) at Lound 07.06.1992.
1994   A 1st summer ♂ in song at Idle Stop 22-24.05.1994.
1995   1 (or possibly 2) in song at Eakring 03.06.1995.
1 (probably a 1st summer ♂) in song at Thorndale Plantation 11.06.1995.
1998   A ♂ in song at Clifton Grove 27.06.1998.
2002   1 in song at the Delta, Attenborough NR 02.06.2002.
2005   A ♂ in song Clumber Park 11-22.05.2005.

This intriguing group of records suggests that occasional overshooting migrant birds briefly settle down on territory in Nottinghamshire before moving on in search of mates. May and June are the best months for records of this species in England.

□ **RED-BACKED SHRIKE** *Lanius collurio*

This shrike was formerly a rare summer visitor to Nottinghamshire and bred in the county until 1947. Perhaps surprisingly a further pair nested successfully at Clumber in 1977 raising 3 young. Another pair was present in the early part of the summer at Annesley in 1978. There have been no breeding reports since then and the species has now disappeared from England as a regular breeding bird.

Apart from the birds listed above only 8 others have reached the county since 1974:

1975   A ♂ at Kirton Priors 22.04.1975.
1976   A 1st winter bird at Attenborough NR 10.10.1976.
1977   A ♀ at Barton 15.07.1977.
1978   A ♀ or immature at Serlby 28.09.1978.
1984   A ♀ at Attenborough NR 01.09.1984.
1987   A ♂ at Longdale Lane 10.06.1987.
1994   A ♂ at Lound 28.05.1994.

1998   A ♂ at Budby Common 07.06.1998.
Extreme dates 07.04.1927 (locality unrecorded) - 10.10.1976 (Attenborough NR).

☐ GREAT GREY SHRIKE *Lanius excubitor*

This species has always been a scarce winter visitor (between October and early April) to Nottinghamshire but small numbers were regularly found in the county from the early 1960s. This pattern continued until the mid 1970s with up to 9 birds seen in each winter period and regular sightings from the Clumber - Budby area and from well watched sites in the Trent Valley (particularly Attenborough NR, Colwick CP and Holme Pierrepont). Smaller numbers (typically 1 to 4 birds) were found in the period 1978-1991 but overwintering birds were still regular at Budby. Since then overwintering records have become more unusual and the species is now best described as a rare winter visitor to the county.

Great Grey Shrike - birds recorded in Nottinghamshire 1985-2007 *

Year	1985	1986	1987	1988	1989	1990	1991	1992
Birds	2	2	1	3	3	2	1	1
Year	1993	1994	1995	1996	1997	1998	1999	2000
Birds	0	1	0	1	2	4	1	2
Year	2001	2002	2003	2004	2005	2006	2007	Total
Birds	1	2	1-2	4	0	3	2	39-40

Months	J	F	M	A	M	J	J	A	S	O	N	D
Birds	9	2	3	2	0	0	0	0	0	13	5-6	5

Probable returning birds are excluded from the table above.
The only overwintering birds found since 1991 were at Budby (1998-99 and up to 2 birds 2007-08), Clumber Park (1998-99 and another 2004-05), Long Lane at Farndon (2001-02) and at Forest Town near Mansfield (2002-03). Extreme dates 02.10.1972 (Attenborough NR) - 21.04.2006 (Brinsley).
See NBAR 2001 pp.119-121.

□ **WOODCHAT SHRIKE** *Lanius senator*
There is one Victorian county record of this southern European shrike - a ♂ shot at Buck Gates on the west side of Thoresby Park in May 1859.

□ **EURASIAN JAY** *Garrulus glandarius rufitergum*
The Eurasian Jay is a reasonably common resident bird in Nottinghamshire. It is prone to autumn irruptions (for example in 1983, 1996, 2003 and perhaps 1993) which presumably involve continental birds of the race *G.g.glandarius* moving into Britain as a result of failed acorn crops further east. During such irruptions, birds turn up in unusual areas with - for example - 35 in gardens at Mansfield 26-27.09.1983 and small flocks are also recorded. The largest site count recorded for 1974-2007 is 31 birds at Clumber Park 22.11.1981.

□ **MAGPIE** *Pica pica*
A common resident which has increased since the 1970s and spread into many urban and suburban areas. The increase can be seen by comparing the annual site maxima published in the Nottinghamshire Birdwatchers Annual Reports.

Magpie - maximum site counts in Nottinghamshire 1975-2007

Years	1975-79	1980-84	1985-89	1990-94	1995-99	2000-04	2005-07
Site maxima	14	44	52	66	100	100	80

There are three counts of 100 birds for the period - at Attenborough NR in 1997, at Lings Bar roundabout 08.02.1998 and at Wilwell Cutting 12.02.2003.

☐ **NUTCRACKER** *Nucifraga caryocatactes*
There are 2 old records of this eruptive European species - 1 at Ramsdale in the winter of 1871 and 2 shot at Clumber Park in 1883. Sadly Nottinghamshire missed out on the large invasion of this species into Britain in 1968-69.

☐ **WESTERN JACKDAW** *Corvus monedula*
The Western Jackdaw is a common resident species in suburban and rural areas but generally avoids the centre of the city of Nottingham and other large towns in the county.
Outside the breeding season Western Jackdaws form significant roosting and pre-roosting flocks with gatherings of up to 4000 birds at various sites in Nottinghamshire. The largest count for the period 1974-2007 was an exceptional estimate of 6000 birds heading to roost at Hoveringham 29.12.2000.

☐ **ROOK** *Corvus frugilegus*
The Rook is still a common resident bird in Nottinghamshire but there has been a long-term decline in the number of pairs breeding in the county since the 1960s, presumably as a result of changes to farming practices. This decline is shown in the 7 surveys which have taken place since 1928:

Rook - surveys of pairs nesting in Nottinghamshire 1928-1996

Years	1928	1932	1944	1958	1962	1975	1996
Number of nests	6576	6113	10306	17028	10609	8726	5837

311 rookeries were located in 1975 but only 127 were found in the 1996 survey with most nesting birds in the east and north of the county. In both the 1975 and 1996 surveys only one rookery was found with over 200 nests and the largest county rookery is probably at Waterloo Wood near Staunton Hall where 350 nests were counted in 1999. 720 nests were also counted around Staunton in the Vale in 2001.
Large roosts and pre-roost flocks of up to 8000 birds have been recorded in several areas since 1974 - for example in the Idle Valley in the mid 1990s and at Hoveringham and Wollaton Park in Nottingham.
See NBAR 1997 pp.133-137.

☐ **CARRION CROW** *Corvus corone*
A common and widespread resident in most habitats, with pairs breeding throughout the county. The Carrion Crow is less gregarious than other crow species but flocks of 100 to 200 birds are not uncommon and larger gatherings are occasionally recorded.

Carrion Crow - maximum site counts in Nottinghamshire 1975-2007

Years	1975-79	1980-84	1985-89	1990-94	1995-99	2000-04	2005-07
Site maxima	c200	300	c180	1000	700	400	500

1000 in the mixed corvid roost at Wollaton Park 22.10.1990 is the highest recorded count for 1974-2007.

◻ HOODED CROW *Corvus cornix*

This species was a regular winter visitor to Nottinghamshire before the First World War with 300 birds shot on one estate in the winter of 1905-06. Fewer birds were recorded in the mid twentieth century but small numbers wintered in the county through to 1976. Since then the Hooded Crow has disappeared from Nottinghamshire as it has from other parts of England, presumably as a result of changes in migration patterns in continental breeding birds. Only 13 birds have been found in the last 28 years:

Hooded Crow - birds recorded in Nottinghamshire 1980-2007

Years	1980-84	1985-89	1990-94	1995-99	2000-04	2005-07	Total
Birds	8	4	1	0	0	0	13

The most recent record is of a single bird at Beeston Rylands 27.12.1990.

◻ COMMON RAVEN *Corvus corax*

Historically the Common Raven was a rare vagrant to Nottinghamshire and there were just 15 records of 17 birds between 1850 and December 1968. The next record was a bird bearing a Welsh ring which was found dead at Haywood Oaks on 22.10.1983, perhaps the first sign of an eastern expansion in the English range of this species in the late twentieth century. As a result of this expansion there are now breeding pairs close to the county boundary in both Derbyshire and Leicestershire.

After 1983 there were 8 records of 1 or 2 birds in 6 years between 1987 and 1998 including 2 which were present at Carburton and Welbeck 28.04-07.06.1996. Birds also summered in the same area in 1999 and in subsequent years and most recent sightings have come from the Dukeries and surrounding areas with a maximum of 6 birds near Ladysmith Wood at Budby on 22.07.2003. In recent years there have also been an increasing number of records from the western and south eastern fringes of the county, for example at Annesley Pit Top where there were 13 records in 2005.

Breeding has never been proved in Nottinghamshire but there were sightings of Common Raven at as many as 19 sites in 2007 which suggests that birds may colonise the county in the next few years.

◻ COMMON STARLING *Sturnus vulgaris*

A common but rapidly declining breeding species and winter visitor. This decline is starkly demonstrated by the fall in birds wintering in Nottinghamshire. The largest winter roost in the county was of an estimated 700000 - 1000000 at Bingham in March 1963 and large six figure roosts were recorded several times in the county through to 1982, particularly in the Mansfield area. These included 350000 roosting at Berry Hill, Mansfield in November-December 1979 - the largest count for 1974-2007. Between 1982 and 1996 roosting numbers were much reduced although there were several counts in the range 50000 - 70000 birds in the East Leake - Gotham area in the south of the county.

In the last decade maximum flocks have been far smaller as set out below:

Common Starling - maximum site counts in Nottinghamshire 1998-2007

Years	1998	1999	2000	2001	2002	2003	2004	2005	2006	2007
Site Maxima	4290	2000	5000	5500	6000	2500	5000	9000	3000	3500

See NBAR 1993 pp.95-100.

□ ROSE-COLOURED STARLING *Sturnus roseus*

This eruptive east European vagrant has reached Nottinghamshire at least 6 times. 2 occurred in the 19th century with a ♂ shot at Eastwood in October 1851 and another bird shot at Ramsdale in September 1856. Another 4 were recorded between 1947 and 1959 with adults at Wilford 16.03.1947, near Attenborough 25.01.1948, Watnall 01.09.1958 and Woodthorpe in mid October 1959.

No birds have been recorded in the county since then.

□ HOUSE SPARROW *Passer domesticus*

The House Sparrow is still a common resident in Nottinghamshire but one which has undergone a significant decline since at least the mid 1990s. Unfortunately there is little published data to establish the size and extent of the decline in the population of this familiar urban bird:

House Sparrow - maximum site counts in Nottinghamshire 1998-2007

Years	98	99	00	01	02	03	04	05	06	07
Site maxima	100	300	124	100+	144	220	150	120	129	150

300 at Holme Pierrepont on 12.08.1999 is the largest published count for 1974-2007.

□ TREE SPARROW *Passer montanus*

A moderately common resident farmland bird which has declined considerably since the 1970s. However it is still reasonably widespread and is present in significant numbers in areas such as the Trent Valley (between Nottingham and Newark), the Sherwood farmlands and particularly in the Idle Valley, where 73 pairs were found in 1996. The extent of the decline is shown by comparing maximum site counts in 5-year periods:

Tree Sparrow - maximum site counts in Nottinghamshire 1975-2007

Years	1975-79	1980-84	1985-89	1990-94	1995-99	2000-04	2005-07
Site maxima	2000	330	200	350	700	250	200
Counts of 250-2000	8	4	0	2	3	2	0

The highest counts for 1974-2007 have all come from the north of the county - a maximum of 2000 at Blyth in January-February 1979, 1000 at Tiln in mid-January 1978 and 700 at Misson Springs on 15.01.1995.

☐ COMMON CHAFFINCH *Fringilla coelebs*

A very common resident, passage migrant and winter visitor across the whole of Nottinghamshire. Counts of 184 territories at Centre Parcs in 2002 and 142 territories at Lound in the same year underline the strength of the county breeding population. Despite such high breeding numbers relatively few large winter flocks were reported between 1974 and 2007 suggesting that many birds may move further south after breeding in the county:

Common Chaffinch - maximum site counts in Nottinghamshire 1975-2007

Years	1975-79	1980-84	1985-89	1990-94	1995-99	2000-04	2005-07
Site maxima	1000	500	300	450	400	300	400

There are only 3 counts of 500 or more birds since 1974 - 500 at Moorgreen 17.02.1974, 500 at Birklands in February 1981 and 1000 birds at Bothamsall in December 1979.

☐ BRAMBLING *Fringilla montifringilla*

The Brambling is a fairly common winter visitor (primarily between October and April with stragglers in September and May). However wintering numbers fluctuate considerably from year to year as set out below:

Brambling - maximum annual flock sizes in Nottinghamshire 1975-2007

Flock sizes	0-99 birds	100-199 birds	200-299 birds	300-399 birds	400-499 birds	500-999 birds	1000+ birds
Number of years	10	6	7	3	2	2	3

As can be seen from the table above flocks were smaller than 300 birds in 23 of the last 33 years. Much larger influxes have only occurred on two recent occasions. In the winter of 1975-76 flocks of 1000 birds were recorded at Ranby in December 1975 and at Clumber Park in December 1975 and January 1976. Two decades later 1000 were present at Everton Carr on 26.12.1994.

Extreme dates 02.09.1984 (Edwinstowe) - 11.05.1986 (Clumber). In addition there was one exceptional summer record - a ♀ at Bestwood CP 10.07.2004.

☐ EUROPEAN SERIN *Serinus serinus*

There has only been one fully accepted record of this southern European finch in the county - a bird which flew through Bestwood CP on 17.03.1993. *See NBAR 1993 pp.89-90.*

☐ **GREENFINCH** *Carduelis chloris*
A very common resident bird throughout the county.
Large winter flocks gather between November and February and there are several counts of 1000 birds or more during these months for 1974-2007:

Greenfinch - counts of 1000 or more birds in Nottinghamshire 1975-2007

Year	Birds recorded
1974	1000 birds roosted at Clumber 31.12.1974.
1978	1000 birds at Tiln January 1978.
1979	1000 birds at Blyth January - February 1979.
1994	1000 birds in the Idle Valley 28.12.1994.
1995	1000 birds at Everton Carr 07.01.1995.
1998	1200 birds in the Idle Valley 06.11.1998. The highest count for Nottinghamshire 1974-2007.

☐ **GOLDFINCH** *Carduelis carduelis*
The Goldfinch is another common resident species in Nottinghamshire. It was trapped extensively in Victorian times but has increased considerably since that time and particularly from the 1940s onwards.
Winter flocks tend to be smaller than those of Greenfinch and there have only been a handful of flocks of 500 or more birds since 1974.

Goldfinch - counts of 500 or more birds in Nottinghamshire 1975-2007

Year	Birds recorded
1994	At least 500 birds at Everton Carr 30.01.1994. Still c500 birds there 04.03.1994.
	c500-c850 birds at Everton Carr 28-29.12.1994.
1995	1000 at Everton Carr 14.01.1995 rising to 1400 birds there 29.01.1995 (the highest site count for 1974-2007). Further counts of 1200 12.02.1995 and 900 19.02.1995.
1997	500 at Gringley Carr 20.09.1997.

☐ **SISKIN** *Carduelis spinus*
The Siskin is a regular winter visitor in good numbers. Larger flocks have wintered in recent years and this may represent a genuine increase in the wintering population:

Siskin - Maximum site counts in Nottinghamshire 1975-2007

Years	1975-79	1980-84	1985-89	1990-94	1995-99	2000-04	2005-07
Site maxima	150	150	200	400	400	550	213
Counts of 100-550	1	4	7	18	27	36	13

The largest count for the period is of 550 birds on the Osberton estate on 31.12.2002 but the population of birds passing through the county is much larger than this as shown by the ringing of 650 (of perhaps 1000) birds in a Meden Vale garden in 2004.

Breeding was first suspected in 1978 and successfully proved in 1990. By 2004 19-21 pairs were breeding at several sites, centred on the greater Sherwood Forest area.

◻ LINNET *Carduelis cannabina*
A common but declining resident farmland and scrubland species. Numbers have fallen as agricultural intensification has reduced breeding and feeding opportunities. However large flocks of up to 400 birds still gather in many places to feed on winter stubble and seeds and larger numbers are sometimes recorded:

Linnet - Maximum site counts in Nottinghamshire 1975-2007

Years	1975-79	1980-84	1985-89	1990-94	1995-99	2000-04	2005-07
Site maxima	500	225	300	3460	2400	500	800
Counts of 500+	0	0	0	9	9	0	2

The Idle Valley is a particularly important wintering area for this species and in the 1990s there were massive counts there of 3460 birds on 02.10.1993 and 2400 birds on 24.09.1996. Such large gatherings have not been repeated in recent years.

◻ TWITE *Carduelis flavirostris*
The Twite possibly bred in the Mansfield Forest / Ollerton area on one or two occasions in the nineteenth century. However there are no recent breeding records for Nottinghamshire and this species is now a scarce winter visitor to the county.
Records have fallen significantly since the late 1990s with only 19 birds between 1999 and 2003. There were no records at all in 2004 - the first blank year since 1981 - and only 17 birds between 2005 and 2007. This fall in numbers is part of a long-term national decline in wintering Twite.

Twite - birds recorded in Nottinghamshire 1975-2007

Years	1975-79	1980-84	1985-89	1990-94	1995-99	2000-04	2005-06	Total
Birds	245	105	117	268	138	15	17	905

Months	J	F	M	A	M	J	J	A	S	O	N	D
Birds	50	149	119	101	9	0	0	0	29	155	64	229

Records are scattered throughout the county with particular concentrations in the Trent Valley, the Idle Valley, at Lound and - in the 1980s - at Rainworth where a passage of 22 birds moving south was recorded 29.09-22.10.1988. Flock sizes are generally small with only 3 counts in excess of 30 birds in the entire period - an exceptional passage of 132 birds at Bulcote 30.12.1976, 38 at King's Mill Reservoir 10-13.02.1986 and 34 at Misson 21.12.1994.
Extreme dates 01.09.1946 (Colwick CP) - 06.05.1991 (Idle Valley).

☐ **LESSER REDPOLL** *Carduelis cabaret*
Historically the Lesser Redpoll was an uncommon resident bird in Nottinghamshire but numbers increased significantly in the 1960s and 1970s with widespread reports of breeding birds. However, since approximately 1980, breeding numbers have fallen back throughout the English Midlands with very few reports of breeding in Nottinghamshire in recent years.

Winter numbers are reasonably large with a peak count of 1100 at Clumber 07.11.1986, although no flock of more than 400 has been reported since then, perhaps a reflection of reduced breeding numbers in Britain.

Lesser Redpoll – maximum site counts in Nottinghamshire 1975-2007

Years	1975-79	1980-84	1985-89	1990-94	1995-99	2000-04	2005-07
Site maxima	1000	800	1100	400	400	300	148

☐ **COMMON (MEALY) REDPOLL** *Carduelis flammea*
Prior to 1974 this species was a rare vagrant to Nottinghamshire with a handful of 19th and 20th century records.

Since 1974, small numbers have been recorded more regularly with records in 28 of the 34 years to 2007 and annual records since 1995. Normally less than 20 birds are found each year but there was an extremely large invasion in the winter 1995-96 with c200 at Budby and c140 in the Idle Valley in December 1995 and perhaps 600 birds in the county in 1996, with almost 500 at Budby 09.01.1996.

Records since 1996 present a more normal picture of the occurrence of this species:

Common (Mealy) Redpoll – birds recorded in Nottinghamshire 1997-2007

Years	97	98	99	00	01	02	03	04	05	06	07
Birds	17	c64	1	1	2	c21	c7	3	18	20	2

Extreme dates 20.09.2002 (Netherfield) – 25.05.1990 (Sherwood).

☐ **ARCTIC REDPOLL** *Carduelis hornemanni*
There are 5 county records of Arctic Redpoll:
1987   1 at Hills and Holes, Warsop 28-31.01.1987.
1995   At least 1 at Budby 16-c.19.12.1995.
1996   1 at Lound 04-15.01.1996.
        1 at Mapperley Park, Nottingham 21.02.1996.
        Up to 2 at Bestwood CP 13-18.04.1996.
During the large national invasion of this species from November 1995 to May 1996 an exceptional 293 birds were recorded in 26 English counties. *See NBAR 1993 pp.85-88.*

☐ **TWO-BARRED CROSSBILL** *Loxia leucoptera*
A ♂ was shot in the Residence Gardens at Southwell in approximately 1875 and is the only fully acceptable county record of this national rarity.

☐ **COMMON CROSSBILL** *Loxia curvirostra*
An uncommon and irruptive species which was recorded in 31 years of the 34 years 1974-2007 but was often only present in small numbers. However large late spring and summer influxes occurred in at least 7 years within the period - particularly in 1991, 1998 and 2003. The peak count for the period 1974-2007 is of 109 birds at Budby South Forest on 06.04.2003.
Breeding was first proved in the Dukeries in 1967 and was either proved or suspected in 14 years between 1974 and 2007. Breeding records have been concentrated in the mature conifer plantations and mixed woods in the Dukeries and around Clipstone and Budby. In recent years larger numbers have sometimes settled down to breed in the county with at least 12 breeding pairs in 2000 and 11-13 breeding pairs in 2004.

☐ **PARROT CROSSBILL** *Loxia pytyopsittacus*
This species has been recorded once in Nottinghamshire with a flock of 7-9 birds in the Dukeries between 14.12.1990 and 16.02.1991.
This species is prone to periodic invasions outside its normal range and October 1990 - April 1991 witnessed the largest ever influx of Parrot Crossbill into Britain involving over 250 birds.

☐ **PINE GROSBEAK** *Pinocola enucleator*
There is one old record for Nottinghamshire - a ♂ shot near Watnall on 30.10.1890. This was the fourth British record. The specimen is in a private collection in Kent.
See NBAR 2001 p.123.

☐ **COMMON BULLFINCH** *Pyrrhula pyrrhula pileata*
The Common Bullfinch is a fairly common species in Nottinghamshire but there has been a recent decline in breeding numbers in Britain. However this species can be secretive and it is difficult to estimate the extent of any decline in Nottinghamshire in recent years.

Common Bullfinch - number of sites recorded 2000-2006

Years	2000	2001	2002	2003	2004	2005	2006
Number of sites	112	60	68	89	95	76	75

Unlike other finches, Common Bullfinches rarely forms large flocks and the largest count for 1974-2007 is of 50 birds at Newstead Abbey on 16.11.1974.
A putative **Northern Bullfinch** *P. p. pyrrhula* from continental Europe was recorded at Netherfield on 05.12.2004 (with what was possibly another bird at Papplewick 3 days earlier) at a time when record numbers of this race had been recorded in Britain.

◻ HAWFINCH *Coccothraustes coccothraustes*

The Hawfinch has always been an uncommon resident bird in Nottinghamshire. An estimated 60 pairs were present in the county in the early 1970s with significant populations in the Dukeries (c.30 pairs) and a colony at Widmerpool in south Nottinghamshire (c10 pairs).
Numbers have declined since then in several areas and birds disappeared from the south of the county in the 1980s. The Dukeries remains the stronghold for this species with a count of 70 on 05.01.1984 (and an unpublished estimate of 200-250 in January 1988) but even there the population at the traditional site at Clumber Park seems to have declined. However numbers have built up significantly on the Welbeck estate where 65-70 were present 08.02.1999 and where 150 were counted early in 2006.

Hawfinch - maximum annual site counts in Nottinghamshire 1975-2007

Years	1975-79	1980-84	1985-89	1990-94	1995-99	2000-04	2005-07
Maximum annual site counts	12-27	17-70	12-30	17-36	18-70	16-51	14-150

See NBAR 1984 pp.41-43.

□ **LAPLAND BUNTING** *Calcarius lapponicus*
There were only 2 records for this species before 1974 - a bird shot between Ollerton and Edwinstowe in the winter of 1850 and 1 at Everton Carr 15.12.1957.
37-38 birds have been seen since then. This is perhaps a reflection of growing familiarity with the diagnostic features of this rare bunting but it remains a good find.

Lapland Bunting - birds recorded in Nottinghamshire 1975-2007

Years	1975-79	1980-84	1985-89	1990-94	1995-99	2000-04	2005-07	Total
Birds	2	0	7	15-16	8	3	2	37-38

Months	J	F	M	A	M	J	J	A	S	O	N	D
Birds	1	2	1	0	0	0	0	0	2	11	12-13	8

1993 was the best year for Lapland Bunting with 7 or 8 birds.
Records of this species are well scattered around the county with only 9 birds at well watched localities in the Trent Valley.
Extreme dates 29.09.1992 (Idle Valley) - 26.03.1994 (Colwick CP).

□ **SNOW BUNTING** *Plectrophenax nivalis*
A very scarce but reasonably regular winter visitor to Nottinghamshire with records in 27 of the 34 years 1974-2007. Records have included several birds at abandoned pit tops.

Snow Bunting - birds recorded in Nottinghamshire 1975-2007

Years	1975-79	1980-84	1985-89	1990-94	1995-99	2000-04	2005-07	Total
Birds	15	3	28	41	28	5	4	124

Months	J	F	M	A	M	J	J	A	S	O	N	D
Birds	18	18	6	2	0	0	0	0	0	24	35	21

The largest recent count is of 8 birds south at Gringley Carr 22.10.1994 (and at least 8 birds were at Besthorpe in November-December 1961).
Extreme dates 14.09.1961 (Besthorpe) - 21.04.1981 (Misson).

☐ YELLOWHAMMER *Emberiza citrinella*

A common but declining farmland bunting. This attractive species is being squeezed out by agricultural intensification, particularly the loss of hedgerows. However there were still over 400 pairs estimated to be breeding at 78 sites in the county in 2000.

Large flocks build up on winter stubble with 500 in the Idle Valley on 13.01.1991 the largest in the period 1974-2007. In recent years smaller flocks of winter birds have been recorded.

Yellowhammer - maximum site counts in Nottinghamshire 1975-2007

Years	1975-79	1980-84	1985-89	1990-94	1995-99	2000-04	2005-07
Site maxima	200	320	200+	500	200	200	100

☐ CIRL BUNTING *Emberiza cirlus*

The Cirl Bunting has always been a rare species in Nottinghamshire. It was recorded twice before the Second World War with 5 birds shot at Edwinstowe in 1859 and 2 obtained at Bagthorpe on 05.02.1897. There was then something of a revival 1942-47 with several records of singing birds and pairs (principally around Newstead, Blidworth and Oxton) which indicates that a small breeding population was present. However there are only 2 subsequent records of single ♂s at Colwick on 08.01.1949 and 12.03.1953.

The species has declined catastrophically in Britain since then with - for example - no records for Leicestershire/Rutland since 1951 and for Warwickshire since 1959. The nearest breeding population is now in south Devon, the last stronghold of the species in Britain.

☐ ORTOLAN BUNTING *Emberiza hortulana*

There is one 19th century record of this species - a bird allegedly caught on Rock Hill, Mansfield in February 1858. This is an unlikely date as most British records are either in May or between August and October. As a result K.A. Naylor regarded this record as unacceptable. There were 3 further records before 1974 - a ♀ at Nottingham SF 30.09.1945 and 2 there 23.10.1950 and a ♂ at Hoveringham 02.05.1971.

There has been one since:

1997   A 1st winter bird at Hardwick Grange Farm, Clumber Park 30-31.08.1997.

See NBAR 1997 pp.142-143.

☐ LITTLE BUNTING *Emberiza pusilla*

1 in a flock of birds (which also contained 2 Ortolan Buntings) at Nottingham SF on 23.10.1950 currently stands as the sole county record. October is the most frequent month for records of this species in England.

☐ **REED BUNTING** *Emberiza schoeniclus*
Another reasonably common but declining bunting. Breeding records from Attenborough NR illustrate the decline that has taken place:

Reed Bunting - pairs at Attenborough NR, Nottinghamshire 1979-2004

Years	1979	1982	1983	1997	1998	1999	2000	2004
Pairs	61+	46+	c60	15	30	20	14	29

A survey of the species in Nottinghamshire in 2000 revealed 347 singing ♂s at 63 sites and should provide an important baseline for future research into breeding numbers. The Reed Bunting is principally a bird of wetland margins but has also adapted to occupy heathland, farmland and forestry plantations recently - a change which offers hope for the future.

Out of the breeding season this species tends to form moderately sized flocks, often with other wintering finches and buntings:

Reed Bunting - maximum site counts in Nottinghamshire 1975-2007

Years	1975-79	1980-84	1985-89	1990-94	1995-99	2000-04	2005-07
Site maxima	200	100	100+	200+	250	100	100

250 roosted at Rushcliffe CP on 12.02.1999 - the best count for 1974-2007. *See NBAR 2000 pp.143-148 and NBAR 2001 pp.113-118.*

☐ **BLACK-HEADED BUNTING** *Emberiza melanocephala*
There is one historical record of this species - a ♂ shot between Radcliffe-on-Trent and Bingham in June or July 1884. This was the second record for Britain following a questionable record of a bird shot at Brighton, Sussex in November 1868. More recently a ♂ was at Blyth 17-25.05.1966 and there has been one further record since 1974:
1976   A ♂ at Mansfield c.18.05-14.06.1976.
The 1966 and 1976 birds may have been escapes from captivity although many vagrants to Britain have been found in May and June.

☐ **CORN BUNTING** *Milaria calandra*
In the 1970s this species was a fairly common resident, generally distributed in arable habits throughout the county - particularly in the Carrlands in the north, through the Trent Valley and in the Trent Hills and fringes of the Vale of Belvoir in the south. This was still true until perhaps the late 1980s with, for example, 51 singing birds in Rushcliffe District in south Nottinghamshire in 1983. Since that time many of the populations within the county have collapsed and this species is increasingly confined to the northern part of the county (with 68% of all records in Bassetlaw District in 2004):

Corn Bunting - Breeding season records in Nottinghamshire 2000-2005

Years	2000	2001	2002	2003	2004	2005
Number of sites	28	30	15	20	12	14

The population in the north is still relatively healthy. There were 84 singing ♂s in the Idle Valley in 1997 and flocks of between 70 and 200 birds are not unusual. However even there numbers are well down on the early 1990s when there were counts of 500 birds at Misson 29.12.1990 and 500 in the Idle Valley 29.12.1993. There are small residual populations in other parts of the county (for example around Barton-in-Fabis in Rushcliffe district where 110 birds were counted 04.02.2007).

Corn Bunting - maximum site counts in Nottinghamshire 1975-2007

Years	1975-79	1980-84	1985-89	1990-94	1995-99	2000-04	2005-07
Site maxima	300	80	90+	500+	200	200+	250

OTHER RECORDS

The following species have not been admitted to the full Nottinghamshire list for various reasons. Many involve records of birds thought to have originated in captivity and some obvious escapes have been excluded from the accounts which follow. One or two others were rejected because the supporting evidence is weak or the record is otherwise questionable. Only the Snow Goose, Hooded Merganser, Golden Pheasant and Spanish Sparrow are currently on the full British list and none is thought to have been recorded in Nottinghamshire as a truly wild bird. This may also be true of the Ruddy Shelduck records of which are set out in the full list above.

An interesting article on the subject of *Contentious Species in Nottinghamshire* by John Knifton covers several other species (eg Crested Tit and Dartford Warbler) which are alleged to have occurred in the county - *NBAR 1992 pp.93-95.*

□ **BLACK SWAN** *Cygnus atratus*
This Australian species regularly escapes from captivity. There are a number of records for the county.

Black Swan - birds recorded in Nottinghamshire 1975-2005

Years	1975-79	1980-84	1985-89	1990-94	1995-99	2000-04	2005	Total
Birds	0	4	0	1	8	18	5	36

Months	J	F	M	A	M	J	J	A	S	O	N	D	Undated
Birds	4	1	8	6	2	0	0	1	2	4	2	2	4

As yet there is little evidence to suggest that a feral breeding population might one day become established in Nottinghamshire. However, in 2006 at least one pair bred at King's Mill Reservoir and up to 8 birds were present at this site. Up to 3 were also present at Clayworth Common and there were also records from 3 other sites. In 2007 birds were present at 5 localities.

□ **BAR-HEADED GOOSE** *Anser indicus*
This southern Asian goose is a regular escape from captivity and up to 8 have been seen in Nottinghamshire annually since 1977. A pair bred at Lound in 1999-2000.

□ **SNOW GOOSE** *Anser caerulescens*
Escaped Snow Geese have been found in many parts of Nottinghamshire since the 1970s. 1986 was the best year for records with 8 escaped birds at large in the county. There is no evidence that truly wild birds have occurred in Nottinghamshire although a white morph bird went west over Girton and Clumber Park on 14.01.2007 with Pink-footed Geese from the East Coast.

□ **MUSCOVY** *Cairina moschata*
This central and South American duck has been recorded as a feral bird in Nottinghamshire since at least 1995. Small numbers are found annually but there is a larger flock of collection birds at Rufford CP which peaked at 16 birds in 1997. A pair bred at Erewash Meadows in 2003 raising 5 young.

□ **WOOD DUCK** *Aix sponsa*
This North American duck has been recorded a number of times in the county with long-stayed released birds at sites such as Rufford CP. However there is no suggestion that a feral breeding population is developing in Nottinghamshire and no birds have been seen in the county since a ♂ was at Attenborough NR between September 1999 and April 2000.

Wood Duck - birds recorded in Nottinghamshire 1975-2007

Years	1975-79	1980-84	1985-89	1990-94	1995-99	2000-04	2005-2007	Total
Birds	2	2	6	4	7	0	0	21

□ **FALCATED TEAL** *Anas falcata*
There are 2 records of presumed escaped birds in Nottinghamshire:
1985   A ♂ at Lound 13-26.01.1985.
1998   A ♂ at Holme Pierrepont and Colwick CP January-March and June 1998.

□ **MARBLED TEAL** *Marmaronetta angustirostris*
There are 2 records of escaped birds in the county:
1982   1 at Hoveringham 25-31.10.1982.
1996   1 at South Muskham 26.08-15.09.1996 (with gaps).

□ **BAER'S POCHARD** *Aythya baeri*
One old record of a bird shot on the River Trent near Beeston in late April 1911. This record is generally presumed to have been an escaped collection bird or possibly a hybrid Ferruginous Duck.

☐ **HOODED MERGANSER** *Lophodytes cucullatus*
There are 2 records of this North American species for Nottinghamshire:
1996   A ♀ visited Barnstone, Hoveringham, Burton Meadows, Gunthorpe and Colwick CP intermittently from 16.11.1996 to 26.04.1998.
2002   A ♀ at Holme Pierrepont 02.04.2002.
The Hooded Merganser is common in captivity and both birds presumably originated in a collection.

☐ **GOLDEN PHEASANT** *Chrysolophus pictus*
There are 13 records of Golden Pheasant for Nottinghamshire between 1975 and 1998 involving 14 or 15 different birds.

Golden Pheasant - birds recorded in Nottinghamshire 1975-2007

Years	1975-79	1980-84	1985-89	1990-94	1995-99	2000-04	2005-07	Total
Birds	3	4-5	3	1	3	0	0	14-15

It is likely that all of these records involved recently released birds rather than wanders from established feral populations elsewhere in England.

☐ **WHITE PELICAN** *Pelicanus onocratalus*
There was one certain record for the county before 1974 - at Holme Pierrepont and Lowdham 11-16.09.1973.
Two other birds have been found since 1974:
1975   1 at Holme Pierrepont and Attenborough NR 17-18.07.1975.
2006   1 at Lenton 16.05.2006.
Another unidentified Pelican was at Wilford, Holme Pierrepont and Burton Joyce 01-04.11.1981.

☐ **GREATER FLAMINGO** *Phoenicopterus ruber*
There are 2 certain records for Nottinghamshire:
1981   1 at Flintham and Hoveringham 04.10-04.11.1981.
1993   1 at Lound 23.07.1993.
Unidentified Flamingos were seen at Besthorpe and Girton 21-24.10.1967, at Attenborough NR 25.03.1972, in west Nottinghamshire October-November 1973 and in the Erewash Valley 12.10.1976.
There are also 3 records of escaped **Chilean Flamingos** *P. chilensis* in the county since 1971 - at various localities August-December 1971, at South Muskham 15.05.1990 and Attenborough NR 30.11-01.12.1991.

☐ **RED-TAILED HAWK** *Buteo jamaicensis*
An escaped bird was at Barton-in-Fabis in 2006-2007 and was apparently paired with a Common Buzzard. However, an earlier bird shot between Nottingham and Newstead in the autumn of 1850 is accepted by some authors as a genuine transatlantic vagrant. Unfortunately the specimen has now been lost and the record has been assessed and rejected by the BOU.
See NBAR 1991 pp.45-47 for a discussion of this interesting record.

☐ **PURPLE SWAMP-HEN** *Porphyrio porphyrio*
One county record - a bird at Attenborough NR 25.08-14.10.1978.

☐ **EAGLE OWL** *Bubo bubo*
There is an old record of a bird shot in Nottinghamshire in 1908 and there are a handful of recent published records of escaped birds:
1991   1 at Wollaton Park and Mapperley 23.02-06.04.1991.
           The same or another bird at Hoveringham 31.08-02.09.1991.
1992   1 at Gunthorpe 09.06.1992.
1995   1 at Clifton 01-03.04.1995.
2007   1 at Scrooby 14.02.2007.
A number of escaped Eagle Owls are currently breeding in Britain and further records might be expected.

☐ **BLACK WOODPECKER** Dryocopus martius
There are currently no accepted records of this species in Britain. However there are a few intriguing claims of Black Woodpecker and there are 2 such old records for Nottinghamshire - 2 birds said to have been shot near Nottingham before 1840 and a sighting at Park Hall near Mansfield in late December 1907.

☐ **SPANISH SPARROW** *Passer hispaniolensis*
A dubious record of a ♂ bird claimed to have been taken on the River Trent at Wilford in the autumn of 1900 predates any currently accepted record for Britain.

☐ **NORTHERN CARDINAL** *Cardinalis cardinalis*
A bird which visited a bird table at Fountaindale from November 1973 to March 1974 was an unlikely transatlantic vagrant. It was presumably an escaped cage bird.

☐ **RED-HEADED BUNTING** *Emberiza bruniceps*
There are 2 records of this cage bird for Nottinghamshire. The first was a bird at Attenborough NR 04.05.1968.
One other bird has been found since 1974:
1976   A ♂ in the Trent Valley at Attenborough NR, Holme Pierrepont and
           Gunthorpe 19.04-18.05.1976.

In addition an Albatross (*Thalassarche* species), perhaps a **BLACK-BROWED ALBATROSS** *Thalassarche melanophris* or a **YELLOW-NOSED ALBATROSS** *T. chlororhynchos* was apparently shot on or near the Lincolnshire/Nottinghamshire border at Stockwith near Gainsborough on 25.11.1836.

APPENDIX 1: RECORDS OF COUNTY RARITIES IN NOTTINGHAMSHIRE

All species with less than 70 birds recorded in Nottinghamshire since 1970.

Species	1800-1949	1950-1969	1970-1989	1990-2007	Last record
American Wigeon	0	1	1	6	2007
Green-winged Teal	0	0	1	6	2007
Blue-winged Teal	0	0	0	1	2000
Redhead	0	0	0	1	1996
Ring-necked Duck	0	0	3	6	2004
Ferruginous Duck	7	11	2	0	1981
Lesser Scaup	0	0	0	3	1998
Common Eider	1	0	20	38-45	2006
Long-tailed Duck	24+	10	22	19	2007
Velvet Scoter	11	1	2	6	1996
Bufflehead	0	0	0	1	1994
Red Grouse	Several	0	4	0	1988
Black Grouse	Bred	0	0	0	1917
Red-throated Diver	10	12	18	8	2007
Black-throated Diver	2	5	18-19	4	2007
Great Northern Diver	4	1	7	11	2006
Fulmar	0	0	14	12	2006
Manx Shearwater	8	5	20	8	2005
European Storm Petrel	12	3	3	0	1982
Leach's Storm Petrel	7	10	13	5	2003
Northern Gannet	15	14	15	12	2007
Little Bittern	4	0	0	0	1921
Night Heron	1	1	2	0	1977
Squacco Heron	2	0	0	1	1998
Cattle Egret	0	0	0	1	1999
Great White Egret	1	0	1	3	2007
Purple Heron	1	0	2	4	2003
White Stork	4+	0	1	16	2007
Black Stork	1	0	0	2	2002
Glossy Ibis	1-2	0	0	0	1909
Eurasian Spoonbill	3	3	7	10	2006
Black Kite	0	0	1	0	1978
White-tailed Eagle	3	0	0	1	2007
Montagu's Harrier	6	4	3	12	2006
Rough Legged Buzzard	31+	0	8	7	2007
Red-footed Falcon	0	0	1	3	2002
Spotted Crake	34	8	16-17	5	2007
Sora	0	0	0	1	2005
Little Crake	0	0	4	0	1983
Baillon's Crake	3	0	0	0	1922
Corn Crake	Bred	Bred	8	4	2006

Common Crane	1	0	1	14	2007
Little Bustard	2	0	0	0	1866
Great Bustard	1	0	0	0	1906
Black-winged Stilt	9	0	1	0	1974
Stone-Curlew	Bred	1	1	2	2005
Killdeer	0	0	1	0	1981
Kentish Plover	6	5	1	1	1997
American Golden Plover	0	0	1	5	2001
Sociable Lapwing	0	0	1	0	1978
Baird's Sandpiper	0	0	0	1	1998
Pectoral Sandpiper	1	2	11	11	2007
Purple Sandpiper	5	1	4	3	2004
Broad-billed Sandpiper	0	1	0	0	1961
Buff-breasted Sandpiper	0	0	3	2	1995
Great Snipe	5	1	3	0	1989
Long-billed Dowitcher	0	0	0	1	1996
Lesser Yellowlegs	1	0	0	1	1995
Solitary Sandpiper	0	1	0	0	1962
Spotted Sandpiper	0	0	0	1	1995
Wilson's Phalarope	0	1	0	0	1961
Red-necked Phalarope	6	3	3	2	1992
Grey Phalarope	9	6	3-4	6	2005
Pomarine Skua	2	0	4	0	1988
Arctic Skua	1	4	16	23	2006
Long-tailed Skua	1	0	0	3	2007
Great Skua	1	1	5	11	2005
Sabine's Gull	0	1	1	2	2004
Ring-billed Gull	0	0	0	2	1996
Caspian Gull	0	0	0	c63	2007
Gull-billed Tern	5	0	0	1	2006
Caspian Tern	1	2	5	3	1999
Roseate Tern	1	1	3	4-5	2004
Whiskered Tern	0	0	0	2	2003
White-winged Black Tern	2	1	0	7	2006
Common Guillemot	Several	0	0	2	2007
Razorbill	2	0	1	2	1995
Little Auk	31	5	5	4	1995
Puffin	4	1	2	2	1992
Pallas's Sandgrouse	c84	0	0	0	1889
Rose-ringed Parakeet	0	1	10	31	2007
Eurasian Scops Owl	0	0	1	0	1973
Egyptian Nightjar	1	0	0	0	1883
Common Nighthawk	0	0	1	0	1971
Alpine Swift	0	0	0	1	1998
Little Swift	0	0	0	1	2001
European Bee-eater	0	0	1	0	1970
European Roller	0	1	0	0	1966

Hoopoe	6	7	20	10	2006
Wryneck	Bred	14	36	15	2006
Short-toed Lark	0	1	0	0	1950
Shore Lark	1	0	4	4	1994
Red-rumped Swallow	0	0	0	6	1996
Richard's Pipit	0	0	2	9	2007
Blyth's Pipit	0	0	0	1	2003
Cedar Waxwing	0	0	0	1	1996
Dipper	7	1	10	11	2005
Bluethroat	0	0	1-2	0	1979
Dusky Thrush	1	0	0	0	1905
Cetti's Warbler	0	0	1	11	2007
Savi's Warbler	0	0	0	3	1997
Marsh Warbler	0	2	2	4	2004
Great Reed Warbler	0	0	1	1	1997
Icterine Warbler	1	0	0	0	1945
Barred Warbler	0	1	1	2	1999
Dusky Warbler	0	0	0	1	2007
Pallas's Leaf Warbler	0	0	0	1	2002
Yellow-browed Warbler	0	0	0	5	2006
Red-breasted Flycatcher	1	1	0	1	1993
Penduline Tit	0	0	0	1	1994
Golden Oriole	10	1	6	9	2005
Red-backed Shrike	Bred	2	11	2	1998
Woodchat Shrike	1	0	0	0	1859
Nutcracker	3	0	0	0	1883
Hooded Crow	WV	WV	WV	1	1990
Rose-coloured Starling	4	2	0	0	1959
European Serin	0	0	0	1	1993
Arctic Redpoll	0	0	1	5	1996
Two-barred Crossbill	1	0	0	0	1875
Parrot Crossbill	0	0	0	7-9	1991
Pine Grosbeak	1	0	0	0	1890
Lapland Bunting	1	1	9	29	2006
Cirl Bunting	23+	1	0	0	1953
Ortolan Bunting	2	2	1	1	1997
Little Bunting	0	1	0	0	1950
Black-headed Bunting	1	1	1	0	1976

APPENDIX 2: FIRST & LAST DATES FOR MIGRANTS IN NOTTINGHAMSHIRE

SUMMER VISITORS & ANNUAL PASSAGE MIGRANTS

SPECIES	FIRST DATE	LAST DATE	OTHER RECORDS
Garganey	24.02.1952 Locality unknown	20.11.1977 Besthorpe	
Common Quail	18.04.2002 Gringley Carr	10.11.1982 Newstead Abbey	Excluding 19th century claims for December & January.
Honey-Buzzard	12.05.1997 Osberton	05.10.2003 Hoveringham	Excluding a 19th century claim at Inkersal Forest 26.04.1858
Osprey	25.03.2006 Misterton Carr & Welbeck	16.11.1889 Locality unknown	
Hobby	06.04.1999 Wollaton Park	31.10.1982 Locality unknown	Excluding 19th century claims for November & December.
Little Ringed Plover	10.02.1961 Nottingham SF	09.11.2006 Attenborough	
Dotterel	06.04.1997 Gringley Carr	13.09.1998 Idle Valley	
Curlew Sandpiper	01.04.1996 Hoveringham	09.11.1991 Lound	
Whimbrel	20.03.1976 West Bridgford	23.10.1960 Nottingham SF	Excluding 19th century claims for November and December
Wood Sandpiper	16.04.1993 Hoveringham	23.11.1962 Holme Pierrepont	1 Nottingham SF & Holme Pierrepont 12.02-16.07.1961
Sandwich Tern	20.03.1958 Toton	28.10.1976 Netherfield	2 Netherfield & Colwick 06.12.1995
Common Tern	24.03.2003 Eakring	14.11.1960 Locality unknown	
Arctic Tern	11.04.1990 Holme Pierrepont	13.11.2005 Hoveringham	
Little Tern	20.04.1993 Hoveringham 20.04.1997 Lound	05.10.1944 Bulcote	Excluding 19th century claims for December.
Black Tern	12.04.1963 Trent Valley	05.11.1984 Stoke Bardolph	

Turtle Dove	01.04.1981 Attenborough	01.11.2004 Caunton	
Common Cuckoo	25.03.1962 Holme Pierrepont	18.11.1956 Westview Farm, Besthorpe	
European Nightjar	21.04.1847 Locality unknown	10.10.1996 Bingham	
Common Swift	03.04.1992 South Muskham	12.11.1952 Locality unknown	
Sand Martin	28.02.1994 Lound	01.12.1960 Widmerpool	
Barn Swallow	11.03.1990 QMC, Nottingham	03.12.1994 Gringley Carr	
House Martin	18.03.1994 Colwick 18.03.2004 Sturton-le-Steeple & Torworth	30.11.1999 Bestwood CP	
Tree Pipit	30.03.1959 Bramcote 30.03.2002 Fiskerton	22.10.1966 Holme Pierrepont 22.10.1993 Colwick	
Yellow Wagtail	10.02.1980 Gunthorpe	26.11.1960 Teversal	1 Netherfield & Holme Pierrepont 11.12.1977-15.01.1978
Common Nightingale	10.04.1961 near Newark	31.08.1976 Cottam	
Common Redstart	27.03.1967 Burton Joyce	26.11.1970 Locality unknown	
Whinchat	06.04.1980 Lound	21.11.2000 Brierley Forest	1 Ratcliffe-on-Soar 22.02.1959
Northern Wheatear	26.02.2002 Attenborough	13.11.1989 Langley Mill	
Ring Ouzel	18.03.2005 Keyworth	26.11.1856 Edwinstowe 26.11.1998 Clifton	1 Old Basford December 1865 1 Blidworth 28.01.1901 1 Colwick 14.01.1942 1 Netherfield 20.01.1960 1 Rufford 28.12.1963 1 Nottingham 03.02-27.03.1996
Grasshopper Warbler	30.03.1989 Colwick & Edwinstowe	01.10.2006 Netherfield	

Sedge Warbler	31.03.1998 Netherfield	07.11.1979 Mansfield	
Reed Warbler	31.03.2005 Holme Pierrepont	03.11.2002 Holme Pierrepont	
Garden Warbler	29.03.1969 Clumber	25.10.1998 Treswell Wood	1 Locality unknown 28.11.1961 1 Colwick 01.12.1991 1 Netherfield 04.12.1969 1 Sherwood 14.12.1980 2 Attenborough 20.12.1969
Lesser Whitethroat	05.04.1994 Attenborough	24.11.1997 Gedling Pit Top	1 Bingham 08.12.1978 1 Blyth 13.01.1979 1 West Bridgford 11.02-01.03.1996
Common Whitethroat	22.03.2002 Lound	13.10.1991 Holme Pierrepont	1 King's Mill Reservoir 28.11.2004 1 Attenborough 10.12.1966 1 Attenborough 22.01.1972
Wood Warbler	03.04.1886 Locality unknown	18.09.1960 Clumber	
Willow Warbler	07.03.1997 Coxmoor GC	12.11.1986 Colwick	1 Misson 30.01-15.03.2005
Spotted Flycatcher	18.04.1977 Norwell	14.10.1981 East Bridgford 14.10.1983 Locality unknown	
Pied Flycatcher	02.04.1960 Fountaindale	06.11.1971 Girton	

Common Sandpiper, **Blackcap** and **Common Chiffchaff** are not included in the table as there are frequent winter records of all three species.

WINTER VISITORS*
*All species with over 70 birds recorded in the county since 1970.

SPECIES	FIRST DATE	LAST DATE	OTHER RECORDS
Bewick's Swan	04.10.1998 Lound	16.05.1987 Lound	
Whooper Swan	20.09.1995 Bennerley Marsh	16.06.2007 Kilvington NL	1 Welbeck 02.07 -08.10.1949
Bean Goose	19.12.1981 Barton Fields	04.03.1982 Attenborough	Excluding feral or escaped birds
Pink-footed Goose	11.09.2005 Annesley Pit Top	13.05.1995 Lound	Excluding feral or escaped birds
White-fronted Goose	02.10.2007 Lambley	02.05.2000 Hoveringham	Excluding feral or escaped birds
Barnacle Goose	September 1869 Ramsdale	26.03.1996 Girton	Excluding feral or escaped birds
Brent Goose	11.10.1888 River Erewash & River Trent 11.10.1992 Stoke Bardolph	Early June 1992 Lound	
Smew	23.10.1999 Lound	19.04.1997 Girton	1♀ Attenborough 21.06.1969
Red-breasted Merganser	20.09.2001 Hoveringham	25.05.1989 South Muskham	1♂ Mattersey 09.06.1971 & at Holme Pierrepont 28.06.1971
Red-necked Grebe	18.09.1988 Hoveringham	04.06.1979 Misson	1 Gunthorpe 01.07.1988 1 Lound 20.08.1997
Slavonian Grebe	12.09.1975 Colwick	03.05.1997 Holme Pierrepont	
Eurasian Bittern	05.08.2002 Out Ings	17.04.1968 Torworth	Serlby breeding season 1956-57 1 Attenborough 16.07.1979 1 Lound 12.05-31.12.1993
Jack Snipe	08.08.1976 Colwick	01.06.1969 Torworth	
Iceland Gull	23.11.1973 Mansfield	02.05.1970 Attenborough 02.05.1997 Lound	
Glaucous Gull	15.10.1991 Rainworth	15.05.1993 Lound	

Water Pipit	06.09.1944 near Netherfield	18.05.1984 Attenborough	
Rock Pipit	10.09.1993 Lound	04.06.1994 Colwick	
Waxwing	09.10.2004 Kirton	11.05.2005 Carrington	1 Collingham 18-19.08.1971
Fieldfare	11.08.1975 Colwick	14.06.1995 Lound	4 Clumber 09.07.1977
Redwing	04.09.1978 Mansey Common	30.05.1992 Bilborough	1 Ashfield Pit Top 18-20.06.1998 1 Mapperley 24.07.1984 1 Collingham 10-12.08.1990
Bearded Tit	02.10.1977 Holme Pierrepont 02.10.1983 Holme Pierrepont	23.04.1977 Trent Valley	1♂ Besthorpe 02.06-20.07.1994 2 King's Mill 21.08.2007
Great Grey Shrike	02.10.1972 Attenborough	21.04.2006 Brinsley	
Hooded Crow	02.10.1872 Ramsdale	29.04.1988 Arnold	1 Ruddington 08.06.1985
Brambling	02.09.1984 Edwinstowe	11.05.1986 Clumber	1♀ Bestwood CP 10.07.2004
Twite	01.09.1946 Colwick	06.05.1991 Idle Valley	Possibly bred 19th century
Common Redpoll	20.09.2002 Netherfield	25.05.1990 Sherwood	
Snow Bunting	14.09.1961 Besthorpe	21.04.1981 Misson	

BIBLIOGRAPHY

GENERAL

Andy Brown and Phil Grice *Birds in England* London 2004.
JN Dymond, PA Fraser and SJM Gantlett *Rare Birds in Britain and Ireland* Calton 1989.
LGR Evans *Rare Birds in Britain 1800-1990* Amersham 1994.
RA Frost *Birds of Derbyshire* Buxton 1978.
David Wingfield Gibbons, James B Reid and Robert A Chapman *The New Atlas of Breeding Birds in Britain and Ireland 1988-1991* London 1993.
Simon Harrap and Nigel Redman *Where to watch birds in Britain* London 2003.
Graham and Jane Harrison *The New Birds of the West Midlands* Studley 2005.
Ron Hickling *Birds in Leicestershire and Rutland* Leicester 1978.
Simon Holloway *The Historical Atlas of Breeding Birds in Britain and Ireland: 1875-1900* London 1996.
Peter Lack *The Atlas of Wintering Birds in Britain and Ireland* Calton 1986.
Chris Mead *The state of the nations birds* Stowmarket 2000.
KA Naylor *A reference manual of Rare Birds in Britain and Ireland - Volume 1* Nottingham 1996.
Philip Palmer *Firsts for Britain and Ireland 1600-1999* Chelmsford 2000.
Adrian Pitches and Tim Cleaves *Birds new to Britain and Ireland 1980-2004* London 2005.
JTR Sharrock *Scarce Migrant Birds in Britain and Ireland* Berkhamsted 1974.
JTR Sharrock *The Atlas of Breeding Birds in Britain and Ireland* Calton 1976.
JTR Sharrock and PJ Grant *Birds new to Britain and Ireland* Calton 1982.
JTR and EM Sharrock *Rare Birds in Britain and Ireland* Berkhamsted 1976.
Andrew Wilson and Russell Slack *Rare and Scarce Birds in Yorkshire* Guildford 1996.

PERIODICALS & COUNTY BIRD REPORTS

Birding World Vol 1/1 (January 1988) to date.
British Birds Vol 72 (1979) to date.
Derbyshire Annual Reports 1974 - to date.
Leicestershire and Rutland Annual Reports 1974 - to date.
Northamptonshire Annual Reports 1974 - to date.
Twitching Vol 1/1 (January 1987) - Vol 1/12 (December 1987).
West Midland Bird Club (Staffordshire, Warwickshire, Worcestershire and the West Midlands) Annual Reports 1974 - to date.

NOTTINGHAMSHIRE RECORDS

MC Dennis *The Wildlife of Colwick Park.*
Austen Dobbs *The Birds of Nottinghamshire Past and Present* Newton Abbot, 1975.
Austen Dobbs *The Birds of Clumber Park* Nottingham 1976.
Anthony Irons *Breeding of the Honey Buzzard (Pernis apivorus) in Nottinghamshire* South Normanton 1980.
John Knifton *Rare Birds in Nottinghamshire 1759-1992* Nottingham 1992.
Nottinghamshire Birdwatchers (formerly the Trent Valley Bird Watchers) Annual Reports 1943 to date.
Nottinghamshire Birdwatchers Newsletters March 2003 to date.
KA Naylor *The Rare and Scarce Birds of Nottinghamshire* Nottingham 1996.
Paul Naylor *Annesley Pit Top Wildlife 2004 and 2005.*
WJ Sterland and Joseph Whitaker *Descriptive List of the Birds of Nottinghamshire* Mansfield 1879.
Joseph Whitaker *Birds* pp.156-176 in *The Victorian history of the Counties of England - Nottinghamshire* London 1906.
Joseph Whitaker *Notes on the Birds of Nottinghamshire* Nottingham 1907.

In addition Phil Palmer supplied some unpublished records and the current county recorder Andy Hall clarified some published records.

USEFUL WEBSITES

Records for Nottinghamshire are considered and published by Nottinghamshire Birdwatchers. The society also runs a lively programme of indoor and outdoor meetings.
The addresses of the current County Bird Recorder and Membership Secretary are listed on the Nottinghamshire Birdwatchers website (www.nottsbirders.net).
The Nottinghamshire Birdwatchers website provides useful links to other websites for sites within the County including Attenborough NR, Clifton Grove, Eakring, King's Mill Reservoir, Lound and Netherfield.
The website also provides details of other local bird clubs including the Derbyshire Ornithological Society, the Leicestershire and Rutland Ornithological Society, the Lincolnshire Bird Club and the SK 58 Birders (who cover Sheffield, Rotherham and Worksop).

FULL COUNTY CHECKLIST

Guidelines for the submission of records to Nottinghamshire Birdwatchers Rarities Committee:
+   Species requiring supporting details.
\#   Species for which further details may be requested by the County Recorder.

In addition descriptions are required for all rare sub-species and supporting details may be required for early or late migrants.

BREEDING STATUS

FB   Former Breeder (with the date of the last breeding record).
OB   Occasional Breeder (with the date of the last breeding record).
ERB   Extremely Rare Breeder (1-5 pairs).
RB   Rare Breeder (6-25 pairs).
SB   Scarce Breeder (26-100 pairs).
UCB   Uncommon Breeder (101-750 pairs).
CB   Common Breeder (750+ pairs).
?   Possible Breeding Records.

CURRENT STATUS

VRV   Very rare (10 or fewer county records to 2007 - with details of the number of birds recorded in Nottinghamshire to 2007).
RV   Rare (11-50 county records to 2007).
S   Scarce (51 to 200 county records to 2007).
UC   Uncommon (annual or virtually annual with up to 20 records per year).
FC   Fairly common (likely to be seen during most visits to suitable habitats but usually in relatively small numbers).
C   Common (likely to be seen in reasonable numbers on all visits to suitable habitat).
R   Resident.
WV   Winter Visitor.
SV   Summer Visitor.
PM   Passage Migrant.
*   Indicates that a species was a formerly a regular breeder in the county.

		BREEDING STATUS	STATUS
☐	1. MUTE SWAN	UCB	CR
☐	2. BEWICK'S SWAN		FC/UCWV
☐	3. WHOOPER SWAN		FC/UCWV
☐	4. BEAN GOOSE +		R/SWV
☐	5. PINK-FOOTED GOOSE		CPM
☐	6. WHITE-FRONTED GOOSE +	OB (Feral)	UCPM&WV
☐	7. GREYLAG GOOSE	CB	CR
☐	8. CANADA GOOSE	CB	CR
☐	9. BARNACLE GOOSE # (wild birds only)	ERB (Feral)	VRV
☐	10. BRENT GOOSE (Pale-bellied subspecies+)		SWV
☐	11. EGYPTIAN GOOSE	RB	UCR
☐	12. RUDDY SHELDUCK		VRV (2)
☐	13. COMMON SHELDUCK	SB	FCPM&SV
☐	14. MANDARIN DUCK	OB (2007)	UCR
☐	15. EURASIAN WIGEON	ERB	CWV
☐	16. AMERICAN WIGEON +		VRV (8)
☐	17. GADWALL	SB	FCR&WV
☐	18. EURASIAN TEAL	ERB/RB	CWV
☐	19. GREEN-WINGED TEAL +		VRV (7)
☐	20. MALLARD	CB	CR&WV
☐	21. PINTAIL	OB (1969?)	UCPM&WV
☐	22. GARGANEY #	OB (2004)	UCPM
☐	23. BLUE-WINGED TEAL +		VRV (1)
☐	24. SHOVELER	RB	FCWV
☐	25. RED-CRESTED POCHARD	ERB (Feral)	SSV&WV
☐	26. COMMON POCHARD	RB	CWV
☐	27. REDHEAD +		VRV (1)
☐	28. RING-NECKED DUCK +		VRV (9)
☐	29. FERRUGINOUS DUCK +		RV
☐	30. TUFTED DUCK	UCB	CWV
☐	31. GREATER SCAUP		UCPM&WV
☐	32. LESSER SCAUP +		VRV (3)
☐	33. COMMON EIDER +		RV
☐	34. LONG-TAILED DUCK +		SWV
☐	35. COMMON SCOTER		UCPM
☐	36. VELVET SCOTER +		RV
☐	37. BUFFLEHEAD +		VRV (1)
☐	38. COMMON GOLDENEYE		CWV
☐	39. SMEW		UCWV
☐	40. RED-BREASTED MERGANSER		SWV
☐	41. GOOSANDER		FCWV
☐	42. RUDDY DUCK	SB	FCR
☐	43. RED GROUSE		VRV (Several)
☐	44. BLACK GROUSE +	FB (c1910)	VRV*(0 recent)
☐	45. RED-LEGGED PARTRIDGE	CB	CR
☐	46. GREY PARTRIDGE	CB	FCR
☐	47. COMMON QUAIL	RB/SB	UCSV
☐	48. COMMON PHEASANT	CB	CR
☐	49. RED-THROATED DIVER +		RV
☐	50. BLACK-THROATED DIVER +		RV
☐	51. GREAT NORTHERN DIVER +		RV
☐	52. LITTLE GREBE	UCB	CR
☐	53. GREAT CRESTED GREBE	UCB	CR

	BREEDING STATUS	STATUS
☐ 54. RED-NECKED GREBE +		SWV
☐ 55. SLAVONIAN GREBE +		SWV
☐ 56. BLACK-NECKED GREBE #	ERB/RB	UCPM
☐ 57. FULMAR +		RV
☐ 58. MANX SHEARWATER +		RV
☐ 59. EUROPEAN STORM PETREL +		RV
☐ 60. LEACH'S STORM PETREL +		RV
☐ 61. NORTHERN GANNET +		RV
☐ 62. GREAT CORMORANT	UCB	CR
☐ 63. SHAG #		SWV
☐ 64. EURASIAN BITTERN		SWV
☐ 65. LITTLE BITTERN +		VRV (4)
☐ 66. NIGHT HERON +		VRV (4)
☐ 67. SQUACCO HERON +		VRV (3)
☐ 68. CATTLE EGRET +		VRV (1)
☐ 69. LITTLE EGRET #		UCV
☐ 70. GREAT WHITE EGRET +		VRV (5)
☐ 71. GREY HERON	UCB	CR
☐ 72. PURPLE HERON +		VRV (7)
☐ 73. WHITE STORK #		RV
☐ 74. BLACK STORK +		VRV (3)
☐ 75. GLOSSY IBIS +		VRV (1-2)
☐ 76. EURASIAN SPOONBILL +		RV
☐ 77. HONEY-BUZZARD +	ERB	S/UCSV
☐ 78. BLACK KITE +		VRV (1)
☐ 79. RED KITE #	FB (early 19th cent.)	SWV&SV
☐ 80. WHITE-TAILED EAGLE +		VRV (4)
☐ 81. MARSH HARRIER		UCPM
☐ 82. HEN HARRIER #		UCPM&WV
☐ 83. MONTAGU'S HARRIER +	OB (1955-56?)	RV
☐ 84. NORTHERN GOSHAWK +	ERB	UCR
☐ 85. EURASIAN SPARROWHAWK	UCB	CR
☐ 86. COMMON BUZZARD	SB	FCR
☐ 87. ROUGH LEGGED BUZZARD +		RV
☐ 88. OSPREY #		UCPM
☐ 89. COMMON KESTREL	UCB	CR
☐ 90. RED-FOOTED FALCON +		VRV (4)
☐ 91. MERLIN	FB (19th century)	UCWV&PM
☐ 92. HOBBY	RB	FC/UCSV
☐ 93. PEREGRINE FALCON	RB	FC/UCR
☐ 94. WATER RAIL	ERB/RB	FC/UCWV
☐ 95. SPOTTED CRAKE +	OB (c1871)	RV
☐ 96. SORA +		VRV (1)
☐ 97. LITTLE CRAKE +		VRV (4)
☐ 98. BAILLON'S CRAKE +		VRV (3)
☐ 99. CORN CRAKE +	FB (1968)	VRV*(9 recent)
☐ 100. MOORHEN	CB	CR
☐ 101. COMMON COOT	CB	CR
☐ 102. COMMON CRANE +		RV
☐ 103. LITTLE BUSTARD +		VRV (2)
☐ 104. GREAT BUSTARD +		VRV (1)
☐ 105. OYSTERCATCHER	SB	FCPM
☐ 106. BLACK-WINGED STILT +	OB (1945)	VRV (Several)

		BREEDING STATUS	STATUS
☐	107. AVOCET #		SPM
☐	108. STONE-CURLEW +	FB (1891)	VRV*(4 recent)
☐	109. LITTLE RINGED PLOVER	SB	FCPM&UCSV
☐	110. RINGED PLOVER	RB/SB	FCPM
☐	111. KILLDEER +		VRV (1)
☐	112. KENTISH PLOVER +		RV
☐	113. DOTTEREL #		UCPM
☐	114. AMERICAN GOLDEN PLOVER +		VRV (6)
☐	115. EUROPEAN GOLDEN PLOVER		CPM&WV
☐	116. GREY PLOVER		UCPM
☐	117. SOCIABLE LAPWING +		VRV (1)
☐	118. NORTHERN LAPWING	CB	CR&WV
☐	119. RED KNOT		UCPM
☐	120. SANDERLING		UCPM
☐	121. LITTLE STINT		UCPM
☐	122. TEMMINCK'S STINT #		SPM
☐	123. BAIRD'S SANDPIPER +		VRV (1)
☐	124. PECTORAL SANDPIPER #		RV
☐	125. CURLEW SANDPIPER		UCPM
☐	126. PURPLE SANDPIPER #		RV
☐	127. DUNLIN		FCPM
☐	128. BROAD-BILLED SANDPIPER +		VRV (1)
☐	129. BUFF-BREASTED SANDPIPER +		VRV (5)
☐	130. RUFF		FCPM
☐	131. JACK SNIPE		FC/UCWV
☐	132. COMMON SNIPE	ERB	FCPM&WV
☐	133. GREAT SNIPE +		VRV (9)
☐	134. LONG-BILLED DOWITCHER +		VRV (1)
☐	135. WOODCOCK	UCB	FCR
☐	136. BLACK-TAILED GODWIT		UCPM
☐	137. BAR-TAILED GODWIT		UCPM
☐	138. WHIMBREL		UCPM
☐	139. EURASIAN CURLEW	ERB/RB	FC/UCPM
☐	140. SPOTTED REDSHANK		UCPM
☐	141. COMMON REDSHANK	RB/SB	FCWV&PM
☐	142. GREENSHANK		UCPM
☐	143. LESSER YELLOWLEGS +		VRV (2)
☐	144. SOLITARY SANDPIPER +		VRV (1)
☐	145. GREEN SANDPIPER		FCPM&WV
☐	146. WOOD SANDPIPER		UCPM
☐	147. COMMON SANDPIPER	ERB	FCPM
☐	148. SPOTTED SANDPIPER +		VRV (1)
☐	149. TURNSTONE		UCPM
☐	150. WILSON'S PHALAROPE +		VRV (1)
☐	151. RED-NECKED PHALAROPE +		RV
☐	152. GREY PHALAROPE +		RV
☐	153. POMARINE SKUA +		VRV (6)
☐	154. ARCTIC SKUA +		RV
☐	155. LONG-TAILED SKUA +		VRV (4)
☐	156. GREAT SKUA +		RV
☐	157. MEDITERRANEAN GULL	OB (Hybrid 1995)	UCV
☐	158. LITTLE GULL	OB (Attempt1987)	UCPM
☐	159. SABINE'S GULL +		VRV (4)

	BREEDING STATUS	STATUS
☐ 160. BLACK-HEADED GULL	UCB/CB	CWV&PM
☐ 161. RING-BILLED GULL +		VRV (2)
☐ 162. COMMON GULL	OB (1969)	CWV&PM
☐ 163. LESSER BLACK-BACKED GULL	OB (2007)	CWV&PM
☐ 164. HERRING GULL		CWV&PM
☐ 165. WESTERN YELLOW-LEGGED GULL #		UCPM
☐ 166. CASPIAN GULL +		SWV
☐ 167. ICELAND GULL #		SWV
☐ 168. GLAUCOUS GULL #		SWV
☐ 169. GREAT BLACK-BACKED GULL		CWV
☐ 170. KITTIWAKE		UCPM
☐ 171. GULL-BILLED TERN +		VRV (6)
☐ 172. CASPIAN TERN +		RV
☐ 173. SANDWICH TERN		UCPM
☐ 174. ROSEATE TERN +		VRV (9-10)
☐ 175. COMMON TERN	SB	FCPM&SV
☐ 176. ARCTIC TERN		UCPM
☐ 177. LITTLE TERN		SPM
☐ 178. WHISKERED TERN +		VRV (2)
☐ 179. BLACK TERN	OB (Attempt 1978)	UCPM
☐ 180. WHITE-WINGED BLACK TERN +		VRV (10)
☐ 181. COMMON GUILLEMOT +		VRV (Several)
☐ 182. RAZORBILL +		VRV (5)
☐ 183. LITTLE AUK +		RV
☐ 184. PUFFIN +		VRV (9)
☐ 185. PALLAS'S SANDGROUSE +	OB (1888?)	VRV (Several)
☐ 186. ROCK DOVE	CB	CR
☐ 187. STOCK DOVE	CB	CR
☐ 188. WOOD PIGEON	CB	CR
☐ 189. COLLARED DOVE	CB	CR
☐ 190. TURTLE DOVE	UCB	FCSV
☐ 191. ROSE-RINGED PARAKEET +		RV
☐ 192. COMMON CUCKOO	UCB	FCSV
☐ 193. BARN OWL	UCB	FC/UCR
☐ 194. EURASIAN SCOPS OWL +		VRV (1)
☐ 195. LITTLE OWL	UCB	FCR
☐ 196. TAWNY OWL	UCB	FCR
☐ 197. LONG-EARED OWL	RB/SB	UCR/WV
☐ 198. SHORT-EARED OWL	OB (1999)	UCWV
☐ 199. EUROPEAN NIGHTJAR	SB	UCSV
☐ 200. EGYPTIAN NIGHTJAR +		VRV (1)
☐ 201. COMMON NIGHTHAWK +		VRV (1)
☐ 202. ALPINE SWIFT +		VRV (1)
☐ 203. COMMON SWIFT	CB	CSV
☐ 204. LITTLE SWIFT +		VRV (1)
☐ 205. COMMON KINGFISHER	SB	FCR
☐ 206. EUROPEAN BEE-EATER +		VRV (1)
☐ 207. EUROPEAN ROLLER +		VRV (1)
☐ 208. HOOPOE +		RV
☐ 209. WRYNECK +	FB (19th century)	SPM*
☐ 210. GREEN WOODPECKER	UCB	CR
☐ 211. GREAT SPOTTED WOODPECKER	UCB	CR
☐ 212. LESSER SPOTTED WOODPECKER	SB/UCB	UCR

	BREEDING STATUS	STATUS
☐ 213. SHORT-TOED LARK +		VRV (1)
☐ 214. WOOD LARK #	SB	UCSV&PM
☐ 215. SKY LARK	CB	CR
☐ 216. SHORE LARK +		VRV (9)
☐ 217. SAND MARTIN	CB	CSV
☐ 218. BARN SWALLOW	CB	CSV
☐ 219. RED-RUMPED SWALLOW +		VRV (6)
☐ 220. HOUSE MARTIN	CB	CSV
☐ 221. RICHARD'S PIPIT +		RV
☐ 222. BLYTH'S PIPIT +		VRV (1)
☐ 223. TREE PIPIT	UCB	FCSV
☐ 224. MEADOW PIPIT	UCB/CB	CR&PM
☐ 225. WATER PIPIT #		SPM&WV
☐ 226. ROCK PIPIT #		SPM&WV
☐ 227. YELLOW WAGTAIL	UCB/CB	CSV
☐ 228. GREY WAGTAIL	SB	FCWV&R
☐ 229. PIED WAGTAIL	CB	CR
☐ 230. WAXWING		UCWV
☐ 231. CEDAR WAXWING +		VRV (1)
☐ 232. DIPPER #		RV
☐ 233. WREN	CB	CR
☐ 234. DUNNOCK	CB	CR
☐ 235. ROBIN	CB	CR
☐ 236. COMMON NIGHTINGALE	ERB/RB	UC/SSV
☐ 237. BLUETHROAT +		VRV (1-2)
☐ 238. BLACK REDSTART	FB (2003)	SPM&SV
☐ 239. COMMON REDSTART	SB	FC/UCSV
☐ 240. WHINCHAT	FB (2007)	FCPM*
☐ 241. COMMON STONECHAT	OB (2006)	FC/UCWV*
☐ 242. NORTHERN WHEATEAR	FB (1970)	FCPM*
☐ 243. RING OUZEL #	OB (1856)	SPM
☐ 244. BLACKBIRD	CB	CR
☐ 245. DUSKY THRUSH +		VRV (1)
☐ 246. FIELDFARE	OB (1984?)	CWV
☐ 247. SONG THRUSH	CB	CR
☐ 248. REDWING		CWV
☐ 249. MISTLE THRUSH	CB	CR
☐ 250. CETTI'S WARBLER +	OB (2007)	RV
☐ 251. GRASSHOPPER WARBLER	SB	FC/UCSV
☐ 252. SAVI'S WARBLER +		VRV (3)
☐ 253. SEDGE WARBLER	UCB/CB	CSV
☐ 254. MARSH WARBLER +	OB (1969)	VRV (8)
☐ 255. REED WARBLER	UCB/CB	FCSV
☐ 256. GREAT REED WARBLER +		VRV (2)
☐ 257. ICTERINE WARBLER +		VRV (1)
☐ 258. BLACKCAP	CB	CSV
☐ 259. GARDEN WARBLER	UCB/CB	FCSV
☐ 260. BARRED WARBLER +		VRV (4)
☐ 261. LESSER WHITETHROAT	UCB	FCSV
☐ 262. COMMON WHITETHROAT	CB	CSV
☐ 263. PALLAS'S LEAF WARBLER +		VRV (1)
☐ 264. YELLOW-BROWED WARBLER +		VRV (5)
☐ 265. WOOD WARBLER	FB (1996)	SPM*

	BREEDING STATUS	STATUS
☐ 266. COMMON CHIFFCHAFF	CB	CSV
☐ 267. WILLOW WARBLER	CB	CSV
☐ 268. DUSKY WARBLER		VRV (1)
☐ 269. GOLDCREST	CB	CR
☐ 270. FIRECREST #		SWV&SV
☐ 271. SPOTTED FLYCATCHER	UCB	FCSV
☐ 272. RED-BREASTED FLYCATCHER +		VRV (3)
☐ 273. PIED FLYCATCHER	OB (1998)	SPM
☐ 274. BEARDED TIT #		RV
☐ 275. LONG-TAILED TIT	CB	CR
☐ 276. MARSH TIT	UCB	FCR
☐ 277. WILLOW TIT	UCB	FCR
☐ 278. COAL TIT	CB	CR
☐ 279. BLUE TIT	CB	CR
☐ 280. GREAT TIT	CB	CR
☐ 281. EUROPEAN NUTHATCH	UCB/CB	CR
☐ 282. EURASIAN TREECREEPER	UCB/CB	CR
☐ 283. PENDULINE TIT +		VRV (1)
☐ 284. GOLDEN ORIOLE +		RV
☐ 285. RED-BACKED SHRIKE +	FB (1977)	RV*
☐ 286. GREAT GREY SHRIKE #		SWV/PM
☐ 287. WOODCHAT SHRIKE +		VRV (1)
☐ 288. EURASIAN JAY	UCB/CB	CR
☐ 289. MAGPIE	CB	CR
☐ 290. NUTCRACKER +		VRV (3)
☐ 291. WESTERN JACKDAW	CB	CR
☐ 292. ROOK	CB	CR
☐ 293. CARRION CROW	CB	CR
☐ 294. HOODED CROW		RV
☐ 295. COMMON RAVEN #		SSV&WV
☐ 296. COMMON STARLING	CB	CR
☐ 297. ROSE-COLOURED STARLING +		VRV (6)
☐ 298. HOUSE SPARROW	CB	CR
☐ 299. TREE SPARROW	UCB/CB	FCR
☐ 300. COMMON CHAFFINCH	CB	CR
☐ 301. BRAMBLING		FCWV
☐ 302. EUROPEAN SERIN +		VRV (1)
☐ 303. GREENFINCH	CB	CR
☐ 304. GOLDFINCH	CB	CR
☐ 305. SISKIN	SB	FCWV&PM
☐ 306. LINNET	CB	CR
☐ 307. TWITE +	FB (19th century)	UCPM&WV
☐ 308. LESSER REDPOLL	RB/SB	FCWV&PM
☐ 309. COMMON (MEALY) REDPOLL #		SWV
☐ 310. ARCTIC REDPOLL +		VRV (6)
☐ 311. TWO-BARRED CROSSBILL +		VRV (1)
☐ 312. COMMON CROSSBILL	RB	UCR
☐ 313. PARROT CROSSBILL +		VRV (7-9)
☐ 314. PINE GROSBEAK +		VRV (1)
☐ 315. COMMON BULLFINCH	UCB/CB	FCR
☐ 316. HAWFINCH	SB	UCR
☐ 317. LAPLAND BUNTING +		RV
☐ 318. SNOW BUNTING +		SWV&PM

		BREEDING STATUS	STATUS
☐ 319.	YELLOWHAMMER	UCB/CB	CR
☐ 320.	CIRL BUNTING +	FB (1940s)	RV
☐ 321.	ORTOLAN BUNTING +		VRV (6)
☐ 322.	LITTLE BUNTING +		VRV (1)
☐ 323.	REED BUNTING	UCB/CB	CR
☐ 324.	BLACK-HEADED BUNTING +		VRV (3)
☐ 325.	CORN BUNTING	SB/UCB	FC/UCR

INDEX

AUKS	86-87	NIGHTJARS	92
BEE-EATER & ROLLER	93	NUTHATCH	119
BITTERNS	45	ORIOLE	119-120
BUNTINGS	132-135	OSPREY	53-54
BUSTARDS	59-60	OWLS	90-92
BUZZARDS	52-53	PARTRIDGES	36-37
CHATS	104-108	PARAKEET	89
COOT & MOORHEN	58-59	PETRELS	41-43
CORMORANT & SHAG	44-45	PHEASANT	37-38
CRANE	59	PIGEONS & DOVES	88-89
CROWS	122-124	PIPITS	98-100
CUCKOO	90	QUAIL	37
DIPPER	103	RAILS & CRAKES	55-58
DIVERS	38-39	SANDGROUSE	87
DUCKS	25-36	SHELDUCKS	24-25
DUNNOCK	103-104	SHRIKES	120-122
EAGLE	50	SKUAS	76-77
EGRETS	46	SPARROWS	125-126
FALCONS	54-55	SPOONBILL	49
FINCHES	126-132	STARLINGS	124-125
FLYCATCHERS	116-117	STORKS	47-48
GANNET	43	SWALLOWS & MARTINS	97-98
GEESE	20-24	SWANS	19-20
GREBES	39-41	SWIFTS	92-93
GROUSE	36	TERNS	83-86
GULLS	77-83	THRUSHES	108-110
HARRIERS	50-51	TITS	117-119
HAWKS	52	TREECREEPER	119
HERONS	45-47	WADERS	60-75
HONEY-BUZZARD	49	WAGTAILS	100-102
HOOPOE	93-94	WARBLERS	110-116
IBIS	49	WAXWINGS	102-103
KINGFISHER	93	WOODPECKERS	94-95
KITES	49-50	WREN	103
LARKS	95-96	WRYNECK	94